Embedded Systems

Embedded Systems

Analysis and Modeling with SysML, UML and AADL

Edited by
Fabrice Kordon
Jérôme Hugues
Agusti Canals
Alain Dohet

 WILEY

First published 2013 in Great Britain and the United States by ISTE Ltd and John Wiley & Sons, Inc.

ISTE Ltd
27-37 St George's Road
London SW19 4EU
UK

www.iste.co.uk

John Wiley & Sons, Inc.
111 River Street
Hoboken, NJ 07030
USA

www.wiley.com

Library of Congress Control Number: 2013932630

British Library Cataloguing-in-Publication Data
A CIP record for this book is available from the British Library
ISBN: 978-1-84821-500-9

Printed and bound in Great Britain by CPI Group (UK) Ltd., Croydon, Surrey CR0 4YY

Table of Contents

Foreword

I am pleased to be asked to write this preface for two reasons. One was to provide fair evidence in favor of pre-existing prejudices. The other was to see the use of the PACEMAKER System Specification as a common subject for the application of SysML, MARTE, and AADL.

Previously only familiar with SysML and MARTE through presentation and tutorial, I expanded and sharpened my understanding of their adaptability to each user's (company's) needs. From UML, SysML and MARTE come *notations* that neither prescribe nor proscribe design methods or tools. The power of defining one's own semantics comes at the cost of defining one's own semantics.

In the domain of embedded electronic control systems using software, AADL reigns supreme. I serve on SAE International AS-2C standard subcommittee that issues the AADL standard and its annex documents, so please understand my enthusiasm in the context of a true believer. AADL is not suitable for modeling an entire aircraft, just its avionics. Where SysML/MARTE hands-off to AADL for electronics and software architecture should be at the "edge" of the electronics (sensors, actuators, displays). Lowest-level SysML/MARTE components trace to top-level AADL components.

Tracing between SysML and AADL might be done using RDALTE[1] [2] which links requirements to architecture. Currently RDALTE links to AADL, using OSATE2, but it is designed to be architecture agnostic. To easily add linking to SysML/MARTE models is a feature of RDALTE.

---

1 Requirements Development and Analysis Language Tool Environment.
2 See AADL information center at http://www.aadl.info for more details.

More personally, the PACEMAKER System Specification used in this book and many published papers was provided by Boston Scientific for use in research and education as a public service. Its origin was serendipitous.

While attending Formal Methods 2006 held at McMaster University in Hamilton, Ontario, I happened to catch a ride with Jim Woodcock. Jim originated the Verification Grand Challenge problem of a Mondex electronic wallet used as the subject of several papers at the conference. Mondex was a clean, simple problem for which a specification had been written in Z. Given the dearth of conference attendees from industry Jim pounced upon me exhorting a "real-world" problem, complex enough to challenge formal methodologies, but not so large as to preclude use by small, academic teams. I told him I'd try to find such a problem.

Rooting around, I found a concise, system specification of a pacemaker designed in the 1990s that was company confidential. However, the company's confidentiality policy made no mention of publicly releasing confidential documents akin to declassification of secrets. Therefore approval of public release would need to come from upper management. With the assistance of many people in Boston Scientific's Formal Methods Group, the document was converted to LaTeX, stripped of proprietary content and mentions of Boston Scientific products, and repeatedly reviewed. The Software Quality Research Laboratory at McMaster University agreed to host the document and its FAQ[3]. Finally, all approvals were obtained, and the document released, six months after that fateful car ride with Jim.

This book succinctly explains three popular languages used to model systems, each complex subjects themselves, in remarkably few pages. Those that want to understand fundamental concepts and differences of SysML, MARTE, and AADL will find this book's use of a common subject helpful.

Brian R. LARSON

Research Associate, Kansas State University

U.S. Food and Drug Administration Scholar in Residence

March 2013

---

3 http://sqrl.mcmaster.ca/pacemaker.htm

Foreword

In the introduction to the report[1] of the mission "Generic embedded software components", which was entrusted to me in 2010 by the French Minister for Industry, the Secretary of State for Forward Planning and Development of the Digital Economy, and the General Commissioner for Future Investments, there is a reminder that:

> The technologies of embedded systems, embedded software, and microelectronics have the capacity to transform all the objects of the physical world – from the smallest to the largest, from the most simple to the most complex – into digital, intelligent, autonomous and communicative devices. The emergence of the Web of Objects, the connection between the world of the Web and the world of embedded systems, considerably amplifies this revolution.

Indeed, the general deployment of embedded systems significantly changes our environment, has brought about numerous innovations in products and uses, and impacts all industrial activities and services.

Mastering the engineering of embedded systems is, therefore, a key element in industrial competitiveness. However, achieving this mastery is still a complex and costly process because of the breadth of the technological field, the diversity and complexity of requirements and solutions. This is illustrated well by the different

---

Foreword written by Dominique POTIER.

1 Briques génériques du logiciel embarqué, Dominique Potier, October 2010 (see www.ladocumentationfrancaise.fr/rapports-publics).

"views" of this type of engineering proposed by the *embedded systems common technical baseline*[2] or CTB:

– The "system" view describes the actual embedded system, that is its architecture, and all its hardware and software components (processing units, bus and operating system (OS) networks, *middleware*, application software).

– The "design tools" view shows all the design tools (modeling, simulation) used.

– The "lifecycle" view of the product describes the activities and where they are involved in the design, development and implementation of an embedded system.

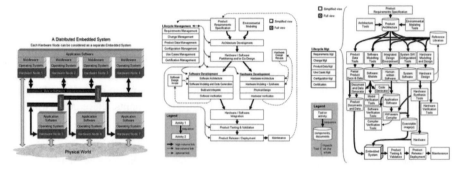

Figure F.1. *"System" views (left), "lifecycle" (center) and "design tools" (right)*

Examination of these different views shows the importance of modeling activities in embedded systems engineering. Therefore, with each level and each element in the "system" view – architecture, processors, bus and networks, software, etc. – correspond, as shown in the "lifecycle" and "design tools" of the product, the models and modeling tools that assist and validate the different stages of development.

This book is, therefore, particularly welcome. It responds to the demand from engineers and students for an approachable book that presents the protocols and principles of modeling embedded systems. We note, in this regard, that there is a particularly strong demand from many engineers with an initial training in electronics who must face a rapid evolution in their activities toward system and software design and modeling. In the report cited above, it is noted that for the SME and research departments in electronics, we are a long way from completing the transformation to embedded software. As noted in the competitiveness and innovation of SME by the electronics (CAPTRONIC) program, these businesses have

---

2 Developed in 2008 by B. Bouyssounouse (VERIMAG), T. Veitshans and C. Lecluse (CEA), the *embedded systems common technical baseline* is a reference, a guide and a knowledge base for embedded systems (see www.embedded-systems-portal.com/CTB).

today too little competence in developing embedded software (a lack of methodology and of tools) and are, therefore, more and more confronted with strict and tough requirements from their clients (instructing parties, system integrators and end users) in terms of robustness and functional reliability.

Another important issue in this book is the process followed to explore the subject. In a pedagogic manner, the presentation of each modeling method is followed by a detailed application of a commonplace case study through which the stages of the activities described in the approaches presented beforehand are clearly demonstrated.

Finally, this book illustrates the dynamism and quality of the research community in France, as much academic as industrial, in terms of modeling languages and tools for embedded systems, and also its role in the bodies standardizing these languages. On this occasion, it is necessary to reaffirm the importance of these standardization activities that allow scientific and technological advances to be recognized and validated by everyone at the international level and that are the reference for developers. In this regard, the MARTE standard (UML *profile for modeling and analysis of real-time embedded systems*), which is presented in part 3 of this book, is a direct result from an initiative launched in 2006 by a group of researchers from THALES and CEA LIST (of whom several are authors of this book) and which resulted in 2009 in the adoption of MARTE standard 1.0 by the Object Management Group (OMG). Moreover, several authors of parts 3 and 4 of this book are also very active in standards bodies for systems modeling language (SysML) and architecture analysis and design language (AADL).

It is my hope that this book meets great success and contributes to creating the next generation of embedded systems designers.

Dominique POTIER

March 2013

Introduction

I.1. The issue

Since the construction of the first embedded system in the 1960s, the *Apollo Guidance Computer* [WIK 12a], embedded systems have continued to spread. They provide a continually increasing number of services and are part of our daily life: elevators, transport (signaling, airplanes, automobiles and railways), telephone services, medicine, energy, industry, etc.

Hence, as we speak more and more of embedded systems, we generally forget their primariy characteristics. The definition given by Wikipedia is as follows [WIK 12c]:

> Embedded computer science represents the software part located in larger systems that are not strictly dedicated to computing. The assembly of software and hardware integrated into a system is called embedded system.

Therefore, a key point in the development of embedded systems is their interaction with their environment and the impact that can result in terms of safety and reliability.

The development of these systems is a difficult problem, which does not yet have a global solution. Of course, airplanes fly, robotic vehicles land on planets millions of kilometers from Earth. However, the development of embedded systems for these purposes still requires a huge amount of energy and work and the use of complex technologies that are difficult to master.

Another difficulty is that systems are plunged into the real world, which is not discrete (as is generally understood in computing), but has a richness of behaviors that sometimes hinder the formulation of simplifying assumptions.

Introduction written by Fabrice KORDON, Jérôme HUGUES, Agusti CANALS and Alain DOHET.

Finally, they are generally autonomous and must face possibly unforeseen situations (e.g., incidents), or even situations that lie outside the initial design assumptions. Let us consider the SPOT [WIK 12b] communication satellites. The project started in 1978 and the first satellite was launched in 1986; the last (sixth-generation) satellites are planned to be launched in 2012 and 2013. SPOT-2, which was originally intended to last for 3 years, was functional for 20 years. This illustrates the difficulty of predicting future uses of technology at the design stage[1].

I.2. Purpose of this book

Having started as a "craft", the construction of embedded systems has become industrialized in line with the increase in the diversity of their applications and use. Methodologies are sketched out, perfected and refined.

The purpose of this book is to indicate the state of the art in the development of embedded systems and, in particular, to concentrate on the modeling and analysis of these systems. These are the key operations that will determine the reliability of future systems.

There are currently three approaches to "model-driven engineering" (MDE[2]): systems modeling language (SysML), unified modeling language/modeling and analysis of real-time and embedded systems (UML/MARTE), and architecture analysis and design language (AADL). All three are the results of international collaborations and focus on different aspects of embedded systems. SysML discusses the systems engineering aspect, whereas UML/MARTE and AADL discuss, in different contexts and with different approaches, the design, analysis and development phases.

Therefore, it is important to present all three approaches in this book in order to provide the reader with a global view of their possibilities. We also demonstrate the contributions of each approach in different stages of the software lifecycle.

I.3. Structure

This book is structured into four parts.

Part 1 comprises two chapters. Chapter 1 presents some general concepts which seem useful to us. Because two of the specification languages are based on UML,

---

1 Remember that computing is a science which is more than 70 years old, if we refer to the creation of the first computer, the Z3 [WIK 12d], by Konrad Zuse in 1941.
2 Section 1.4.1 (Chapter 1) describes this concept.

presenting this notation is useful. Similarly, we also sketch the main issues in developing embedded systems.

Chapter 2 presents the simplified specifications of a case study that is common to the approaches addressed in this book: a pacemaker. The chapter discusses a simplified version of a specification published in 2007 by Boston Scientific [BOS 07]. This case study is a common thread that enables the reader to observe how the different aspects of a system are addressed using different approaches. The interesting point of this case study is that it has also been approached with methods other than those presented in this book, and this allows the reader to compare their expressiveness and the associated analytical tools.

Parts 2–4 each deal with one of the approaches addressed in this book: SysML (Part 2), MARTE (Part 3) and AADL (Part 4). They follow the same structure closely. The first chapter presents the notation associated with the process, and the second chapter discusses the modeling of the case study. These are followed by a chapter dedicated to analysis of the specification. For parts 3 and 4, an additional fourth chapter is dedicated to automatic code generation.

I.4. Acknowledgments

Before closing this introduction, the coordinators wish to thank the various people without whom this book would not have been possible:

– Dominique Potier from the System@tic competitiveness cluster, who wrote Foreword of this book;

– the numerous authors who, dispersed around France, coordinated and exchanged their opinions in order to present a consistent view of the considered approaches;

– the SEE and their editorial committee who encouraged us in our editorial efforts;

– the GDR GPL (Génie de la Programmation et du Logiciel [CNR 12]) of the CNRS which, through its grant, supported coordination meetings for the authors while preparing this book.

I.5. Bibliography

[BOS 07] BOSTON SCIENTIFIC, *PACEMAKER System Specification*, January 2007.

[CNR 12] CNRS, "Page d'accueil du GDR GPL", 2012. Available at gdr-gpl.cnrs.fr.

[WIK 12a] WIKIPEDIA, "Apollo guidance computer", 2012. Available at en.wikipedia.org/wiki/Apollo_Guidance_Computer.

[WIK 12b] WIKIPEDIA, "SPOT (satellite)", 2012. Available at en.wikipedia.org/wiki/SPOT_(satellite).

[WIK 12c] WIKIPEDIA, "Système embarqué", 2012. Available at fr.wikipedia.org/wiki/Systeme_embarque.

[WIK 12d] WIKIPEDIA, "Zuse 3", 2012. Available at en.wikipedia.org/wiki/Z3_(computer).

PART 1

General Concepts

Chapter 1

Elements for the Design
of Embedded Computer Systems

1.1. Introduction

The development of embedded systems is usually done using low-level languages: namely, assembly language and oftentimes C. The main reason for this is the need to elaborate programs that have a small memory footprint because they are usually deployed on very compact architectures[1]. More costly development techniques have been gradually created. These development techniques are mainly based on an in-depth assessment of the requirements, intensive tests as well as very strict development procedures, which ensure a safety standard satisfying the expectations of the general public.

However, these systems that often accomplished critical missions frequently involved very expensive developing strategies, thereby being limited to a specific usage such as space travel, aeronautics, nuclear use and railroad transportation. Once these systems emerged in more "mainstream" industries, the approaches in development had to evolve in a cost-reducing direction. The economic revolution toward offshore development does not facilitate these aspects of viability/safety, but these new development approaches could, in the long run, become very competitive.

Chapter written by Fabrice KORDON, Jérôme HUGUES, Agusti CANALS and Alain DOHET.

1 For example, the probe Pioneer 10, launched in 1972, had only six Kbytes of memory [FIM 74], which did not stop it from transmitting an impressive amount of scientific data regarding Jupiter and its satellites, before becoming the first man-made object to have left the solar system.

The mere use of high-level programming languages is not, in and by itself, the solution. Embedded systems usually contain sophisticated mechanisms and runtime libraries. Because the latter cannot be certified, they are very difficult to use. Take the Ada language as an example of this. It was elaborated in the 1970s with the aim of developing low-cost embedded systems. Some of its features (usually the management of parallelism) have seen limited usage in certain fields, largely due to the code compiled having to be associated with limited runtimes (for instance, not being able to support the parallelism or the dynamic memory allocation).

One solution to maximising software viability and minimizing the cost of embedded computer systems is through the use of model-driven engineering (MDE). It indeed facilitates better interactions between the different languages and models used throughout the design/development cycle.

More particularly, it allows us to rely on dedicated "models", which sometimes facilitate the reasoning process. Thus, engineers can predict certain behavioral aspects of their programs. The transformation techniques that have been developed by this community ensure the link between the different development stages.

These new approaches will not suddenly replace the existing procedures. The actors behind the development of embedded systems (particularly when they carry out critical missions) cannot afford to take any risks. However, these actors have gradually taken an interest in these new techniques and their application in adequate methodological frameworks.

In the long run, the objective is to reach high-performance industrial processes capable of ensuring trustable software, by providing, for instance, co-simulation and formal verification and allowing the target code generation to be validated right from the beginning of the modeling.

It should not surprise us that the three notations presented in this book – systems modeling language (SysML), unified modeling language/modeling and analysis of real-time and embedded systems (UML/MARTE), and architecture analysis and design language (AADL) – are heavily reliant on model engineering. Model engineering is clearly a "hot topic" in the community at the moment.

Chapter outline. This chapter discusses several elements that are important in the development of embedded systems in the context we have just touched upon. We pay particular attention to:

– the modeling activity (section 1.2);

– the presentation of the UML, which serves as the foundation for two of the three notations presented in this book (section 1.3);

– the presentation of the MDE (section 1.4);

– an overview of the analysis techniques and, in particular, those used in this book (section 1.5);

– the methodological aspects of system development (section 1.6).

1.2. System modeling

The modeling operation has been a widespread practice in the history of humanity. It seeks to explain the behavior of a complex system via a model that abstracts it. We can cite the different models designed for explaining celestial movements: the system of the epicycles, the Ptolemaic system, the Tycho Brahe system, the Copernican system, etc. Their objective was to predict the evolution of the planets' position. They represent a process of ongoing fine-tuning of the understanding of a field, a new system replacing the old system when new evidence crops up showing that the old system does not correspond to reality. Modeling has, therefore, been an indispensable tool in experimental sciences for a long time.

In computer science, the big difference lies in the fact that the model does not reproduce a system we are trying to observe in order to understand its behavior. The model is placed at an earlier stage, and allows us to realize whether a solution, which is in the process of being discovered, will respect the properties expected.

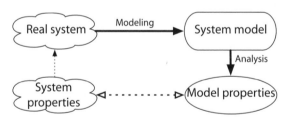

Figure 1.1. *Relations between the system, its model and their properties*

Figure 1.1 represents the relation between different entities of a system at the core of the modeling operation. The engineer models the system while he or she is in the process of designing it. The analysis of this model (via simulation or via more formal methods, see section 1.5.1) allows us to deduce system properties. However, the relation between the properties of the model and those of the system is by no means a trivial issue.

First, the expression of the properties may vary greatly in the two cases. Usually the property of the system will be expressed in a natural language, whereas the property of the model will have the shape of an invariant or will reference quite precisely an entity of the analysis (whose meaning must be known to the users if they wish to understand it). We must thus anticipate elements of traceability between the two.

Second, the necessary abstraction of certain system elements during the modeling process (for instance, changing from discrete types to continuous types) can, when done poorly, result in a model that does not faithfully represent the system. If it is a superset, then a property that has not been verified on the model can be verified on the system; if it is a subset, then a property that has been verified on the model may not be true for the system.

Modeling is thus a challenging operation that must be carried out carefully. More particularly, the modeling choices must be documented properly. In the field of embedded systems, we will introduce several important notions regarding the concept of a model.

Structural or behavioral modeling. The former defines the structure of a system (i.e. the class diagrams or the UML instances) whereas the latter describes its behavior. When examining problematic behaviors, or identifying emerging behaviors, we must reach the second level, which generally presupposes that the modeling of the system structure already exists (at least the identification of the interacting actors).

Open or closed models. An open model describes a system. A closed model describes the system and the environment it interacts with. We can, thus, consider that the open model is "included" in the closed model.

The reason we need to distinguish between the two is that we are able to subject the model of a system to different conditions of execution. Each of them is thus characterized by a dedicated specification of the environment that "closes" the specification. This notion is particularly useful when we study different execution modes: a nominal mode, modes that have been degraded as a consequence of certain conditions, etc.

Notations used in this book. This book deals with embedded systems and only uses the notations that enable us to describe those systems. Out of these, we have selected three. Of the three, SysML [OMG 12a] and MARTE [OMG 12b] are UML profiles, and AADL [SAE 09] is a dedicated notation that integrates, at the same time, the description of the software part and that of the hardware part of a system.

1.3. A brief presentation of UML

The first version of the UML was version 2.8 of the "unified method", which was written by G. Booch and J. Rumbaugh. At the time, in 1995, Y. Jacobson was not yet part of the adventure, and the letter "M" stood for "method".

A year later, the first versions of the UML, namely versions 0.9 and 0.91, were published, which incorporated the work of Y. Jacobson. This time, the "M" stood for "modeling". Indeed, these three "fabulous evangelists", having not agreed upon one standard method, had instead focused on the notation.

UML 1.0 and 1.1 were proposed to the Object Management Group (OMG) in 1997. Then came version 1.2 in 1998, 1.3 (a very significant revision) in 1999, 1.4 in 2001 and, finally, 1.5 in 2002. This sequence of different versions is usually called "UML1.X". This was followed by UML 2.X, whose first stable version (2.0), appeared in 2003. This is a major revision now involving 13 diagrams instead of nine previously. The metamodel was also considerably modified (which is of relevance to tool designers). The notion of the profile was formalized, which has allowed, for example, the emergence of "variants", such as SysML and MARTE, which we will detail in this chapter.

Let us note that the transition from 1.X to 2.X, which was supposed to facilitate the use and efficiency of the language, has given mixed results. The industry and the toolmakers had put a significant amount of time into fully supporting UML 1.X and the required adaptation has delayed the operational use of UML 2.X, which has only recently begun to be used effectively in industry.

We will introduce the main UML characteristics (in its current version in 2012, the 2.4.1 version). The 16 UML diagrams are divided into two classes: static and dynamic diagrams.

Among these 13 diagrams, the four "main" diagrams are considered to be *classes*, *sequence*, *use cases* and *state machines*. They are used in the parts dedicated to SySML and to MARTE but also, systematically, by all the UML users.

UML is a notation. To use it well, we also need a method. Thus, each company has created its own method usually relying on the "unified" method (UM) or the *rational unified process* (RUP) (the two are pretty close). These methods describe when and how to use this or that diagram, how to organize the model as well as the documents and the codes that are generated while still respecting the specification/design processes that are enforced throughout the enterprise such as traceability, configuration management and quality (rules for a correct usage, coding rules, etc.).

1.3.1. *The UML static diagrams*

These diagrams describe the static aspects of a system (relative to their organization). In this category, we may find the following diagrams: *classes*, *composites*, *components*, *deployment*, *object* and *package*.

Class diagrams. They allow us to model the domain (for the sake of simplification, the data and/or the concepts manipulated by the application), and then the application itself. We will thus have the classes of the domain, then the application classes (analysis and design). An average model generally brings about approximately 100.

Composite diagrams. They allow us to decompose a class (in general an application class) into smaller parts. This decomposition enables us to show, for example, which part implements which operation, or how the different parts communicate with each other.

Let us note that there is a second type of composite diagram, which allows us to model *patterns*. Once these patterns are modeled, they can be used in class diagrams.

Components diagrams. They allow us to describe a system via its components and the interactions between components through their interfaces. A component can also be decomposed into subcomponents (just like the classes decompose into smaller parts) by means of a composite diagram.

The difference between a class and a component is the subject of a wide debate. However, our viewpoint can be summed up as follows:

– A class is the basic building block of a software.

– A component is also a building block but of a different level of abstraction; it generally groups together a set of classes, but can also group together other artifacts such as configuration files.

– Classes are interlinked via different relations (heritage, association and composition) and they propose various interfaces.

– Components are interlinked via interfaces.

– Classes can be decomposed into smaller parts.

– Components can only be decomposed into other components (i.e. components having a smaller granularity).

Deployment diagrams. They allow us to model the physical architecture (architectural components) of the application: machines, processes, communication modes, etc. In general, a process is made up of components (which are composed of classes and other artifacts) and distributed to one up to n machines.

Object diagrams. They instantiate a class diagram. A class diagram is a generic model (for instance, the description of the organization of an enterprise) whereas an object diagram is a specific model (for instance, the description of the organization of an enterprise).

In general, we start out by drawing class diagrams and we make sure they concur with the object diagrams. However, several designers prefer to start with the objects (of the specific domain) and then generalize in order to find the classes. The final result is often similar.

Package diagrams. They allow us to organize the model. A *package* may contain *packages*, classes, objects and diagrams. In general, a model is organized in three *packages*: needs analysis (*use cases*), logical architecture (classes) and physical architecture (machines, processes and components). Depending on the method applied in the enterprise, each of these *packages* can be further decomposed.

It is worth noting that a *package* is not a component (neither software nor hardware) but rather a model structuring unit.

1.3.2. *The UML dynamic diagrams*

These diagrams describe the dynamic aspects of a system (i.e. relative to their execution). In this category, we may find the following diagrams: activity, interactions (*sequence, communication, overview* and *timing*), *use cases* and *state machines*.

Activity diagram. This type of diagram (which includes the concepts of parallelism) facilitates the modeling of an algorithm via concatenation of activities/events. It proposes an action language allowing us to specify in more detail all of the algorithmical processing.

Interaction diagrams. (sequence, communication, overview and timing) The sequence diagrams and communication diagrams allow us to model the collaboration between the objects (class instances). These two diagrams are almost equivalent even if they present us with several specificities. The main difference is a visual one: the sequence diagram shows a sequential view, whereas the communication diagram shows a spatial one.

The *overview* diagram allows us to show a concatenation of diagrams in the shape of an algorithm that is similar to the activity diagram. However, the activities and actions are in themselves diagrams. This gives us better readability throughout the specification of complex concatenations.

The *timing* diagram (derived from electronic engineering) facilitates the modeling of behaviors that are sequenced by time events (for instance, the time constraints between different states of several objects).

Use case. They facilitate the description of the system from an external point of view: the actors (roles) using the system as well as the services (*use cases*) offered by the system. The *use cases* are, in general, fine-tuned by activity diagrams and/or sequence diagrams filled in with natural language.

State machines. They model the behavior of an active class (asynchronous events, communication protocols, etc.). They can equally model the behavior of the system (its different states, the transition conditions from one state to another, etc.).

We call an "active class" a class that has an autonomous behavior.

1.4. Model-driven development approaches

The UML 2.0 version appeared in the early years of the 21st century; this version has been progressively integrated in development workshops, which, since then, have started to provide a very rich array of modeling tools. These tools have been developed in accordance with norms that synthesize a number of experiences and industrial expectations in the field of system engineering, thus grouping them together for the first time in years.

Around the same time, the notion of MDE appeared.

Three predominant approaches were born from this concept:

– the OMG approach: model-driven architecture (MDA), based on UML and Meta Object Facility (MOF) (the language that allows us to write metamodels in the OMG world);

– the ECLIPSE approach: *Eclipse Modeling Framework* (EMF) based on ECORE (the language that allows us to write metamodels in the ECLIPSE world);

– the Microsoft approach: tools and concepts based on *domain specific language* (DSL).

Let us note that the approach proposed by the OMG is not well equipped, contrary to the other two. Therefore, it is up to the user to make his or her workshop.

1.4.1. *The concepts*

The MDE mainly consists of using the models in the different phases of the developing cycle of an application. There are three levels that we will consider:

– the requirements or *computation independent model* (CIM);

– the analysis and the design or *platform independent model* (PIM);

– the "pre-code" or *platform specific model* (PSM).

The main objective of the MDE is the elaboration of models that are independent of the technical details of target platforms. This independence must engender, in the long run, the automatic generation of (via model transformation) a large part of the code of the applications, besides spurring a gain in productivity.

Another principle that is highly significant in MDE is the allowing of the transformation of the existing models into the target representation. Thus, whatever the technology used, it is possible to pass very easily from one technology to another, provided that we have the necessary transformation tools. These are based on transformation languages that must follow the *query/view/transformation* (QVT) norm proposed by the OMG.

1.4.2. *The technologies*

It is crucial to know which technology to use. Indeed, for a given domain of application, there are at least two options:

1) writing a metamodel of the domain of application, and then running adapted equipment:

2) writing an UML profile of the domain of application before using an existing "profile" tool.

We will illustrate the advantages and disadvantages of these two techniques using the "model editor" as an example.

In the first case, there are no modeling tools for the chosen technology. The approach consists of writing the metamodel of the technology (i.e. a DSL) and then generating an UML modeling tool (in general 60% of automatic generation with the EMF tools). Once the UML modeling tool is generated, the engineers create a "specific" tool, allowing us to carry out technology models.

In the second case, the choice is made to use an existing UML modeling tool. The approach consists of writing the profile of the technology. Once this profile is created, the engineers create a profiled generic UML tool, allowing us to carry out technology models.

In the first case, the graphic range of the editor proposes concepts such as "buffer" and "task". In the second case, the graphic range of the editor proposes the usual UML concepts such as the "class", but the class can only be stereotyped as a "buffer" or a "task".

The two techniques have their advantages and disadvantages. In the first case, the job engineers have an editor that gives them a job-specific vocabulary, which is an advantage. In contrast, in the second case, they must pass through the notion of "class" before introducing the job concepts, which can be misleading.

In order for these two techniques to be equivalent, we should be able to configure the UML modelers so they present the job concepts without passing through the UML concepts, as we could observe it with SysML or the famous *block*, which is, in fact, a class stereotyped as *block*.

For the time being, the two techniques sit along side-by-side in the industry, belonging to opposite camps, each of them arguing in favor of its advantages.

1.4.3. *The context of the wider field*

The major objective of the models is to facilitate mutual understanding, exchanges and the communal work done by the actors of a project. Their construction and representation must therefore observe certain conventions and rules, which are turned into several norms and standards.

The modeling language determines the manipulated concepts, their semantic and their representation under a textual and graphic form. The variety, of the preoccupations both in the early design/development stages in the domains of application and in the specialties involved (safety, reliability, human factors...) has lead to so many languages that it is impossible to enumerate them all. They can, however, be classified into three large categories:

– The languages with a generic aim: the main language is UML, SysML and MARTE being seen as its derivatives for system engineering and real-time embedded systems.

– The more narrow, specialized languages, which are associated with formal verification methods (automata, Petri nets, B, Lustre, etc.) and allow a mathematical verification of certain expected properties.

– The non-formal specialized languages, connected to certain domains of application, specialties or specific preoccupations, whose terminology they integrate in their construction, as well as incorporating their rules and concepts. The DSL's come from this category. *Workflow* languages, such as the *business process modeling language* (BPML), will be useful in describing the needs for interaction between human beings and the systems.

In the case of embedded computer systems integrated in wider systems, it is necessary to use dedicated architectural frameworks to master both a global consistency and a potential evolution in the future. These frameworks structure and specify how to define the architecture of a system (or the way it will be used by the end user organization). To do that, they define the different required viewpoints for its description, as well as the metamodel necessary to ensure the consistency between the different viewpoints and for carrying out impact analyses.

In comparison to the diagrams prescribed by languages such as UML, the views of the architecture frameworks distinguish themselves through the larger spectrum that they cover (they consider the dimensions that are necessary for the overall governance: enterprise organization and processes, evolution of the capacities, synchronization of the projects, etc.) and they are limited to specifying the type of information that must be provided, this giving a lot of freedom for the actual manner of representation. The NATO Architecture Framework (NAF) is one of the most recent and most advanced architecture frameworks. This is why, it is used beyond its military scope.

Name of the standard	Role of the standard
Fundamental concepts for modeling	
ISO/IEC/IEEE 42010 (2011): System and software engineering – Description of the architecture	Defines the principles that must be respected and the fundamental notions (viewpoints, architecture frameworks, architecture description languages, etc.)
IEEE 1471 (2000): IEEE Recommended Practice for Architectural Description of Software-Intensive Systems	As a reminder: replaced by ISO/IEC/IEEE 42010
Architectural modeling languages	
OMG UML: Unified Modeling Language V2 UML: ISO/IEC 19505 (V2.4.1 - 2012): Information technologies – Unified modeling language	
OMG SysML: System Modeling Language Specification (V1.3 - OMG, 2012/06)	Extension of the UML to system engineering
UML/SPT: Profile for Schedulability, Performance and Time (OMG, 2005)	Extension of the UML for the modeling of real-time systems. Replaced by the MARTE profile
UML/MARTE: Profile for Modeling and Analysis of Real-Time and Embedded Systems (MARTE 1.1 – OMG, 2011/06)	Extension of the UML for the modeling of real-time and embedded systems. Replaces the SPT profile
UML/QFTP: Profile for Modeling Quality of Service and Fault Tolerance Characteristics & Mechanisms (OMG, 2008)	Extension of the UML in order to facilitate the modeling of aspects such as the quality of service and the tolerance to application faults
UML Testing Profile (UTP) (V1.1 – OMG, 2012)	Extension of the UML with concepts relative to the tests
SAE AS 5506 rev. A (2009): Architecture Analysis & Design Language (AADL)	Language for the description of the architecture (software + execution platform) of the real-time embedded systems with critical performance
IEEE Std 1320.2 (1998) – IEEE Standard for Conceptual Modeling Language – Syntax and Semantics for IDEF1X97 (IDEFobject)	IDEF is a family of modeling languages for software and system engineering
OMG BPMN 3.0 (2011): Business Process Modeling and Notation	Graphic notation standard serving to describe the processes of the enterprise (as well as the man–system interaction)
The Open Group – ArchiMate 1.0 Specification (2009)	Enterprise architecture description language
Architectural frameworks	
NATO AC/322-D 0048 (2007): NATO Architecture Framework V3 (NAF V3)	Architecture framework for large (military) systems
ISO 15704, Industrial automation systems – Requirements for enterprise-reference architectures and methodologies	
The Open Group Architecture Framework (TOGAF)	Enterprise architecture
ISO/IEC 10746 (1998): Information technology – Open distributed processing – Reference model (RM-ODP)	Description of distributed information systems

Table 1.1. *The main standards for system modeling*

Table 1.1 gives an overview of the main standards used for system modeling. The main organizations that devise the rules and standards regarding architecture modeling of embedded systems are:

– the International Organization for Standardization (ISO);

– the Institute of Electrical and Electronics Engineers (IEEE), which is an American professional organization; that publishes its own standards via the *IEEE Standards Association*;

– the OMG, an American association whose objective is to standardize and promote the object model in all its forms.

To this list, we may add some so-called "sectoral" actors:

– the International Society of Automotive Engineers (SAE International), which is a world association covering the aerospatial domain as well as the one of terrestrial vehicles, having the aerospace standard (AS) standards and ground vehicle (GV) standards;

– NATO, mainly in the military domain;

– *The Open Group*, for the enterprise information systems;

– the Electronic Industries Alliance (EIA), which stopped its activity in 2011 and has been replaced by five more specialized associations, among which the Governmental Electronics and Information Technology Association (GEIA);

– the European Cooperation for Space Standardization (ECSS), which holds a set of norms dedicated to spatial projects management, organized in three branches: projects management, product insurance and system engineering;

– the Requirements and Technical Concepts for Aviation (RTCA).

1.5. System analysis

For a long time, systems have been validated via tests and simulations whenever the executable models were available. However, this exploratory approach does not guarantee that the set of possible executions is well covered. So-called "shadow areas" within the execution of the system can still exist, and they can hide various faults. This is in particular true in the case of embedded systems. Non-functional constraints (such as energetic consumption and performance) can, indeed, be added to the functional constraints that are already difficult to verify.

To pass beyond the so-called "horizon", which is the limit of traditional simulation and test approaches, we must use formal methods, as shown by numerous studies. Only formal methods guarantee that a property is verified because they rely on mathematical models' existance and enable us to reason on the basis of the specification.

In this domain, there are two families of formal verification techniques: theorem proving and model-checking. Let us briefly recall the principle of the first family that will not be further explained in this chapter because we will not use it. In the following, we present in more detail the model-checking approach, which is used in Chapter 8.

1.5.1. *Formal verification via proving*

The tools and emblematic languages of this family of formal techniques are Coq [AFF 08], Z [ISO 02] or B [ABR 96]. They all have their own success stories such as the behavior verification of metro line 14 in Paris, using method B.

The basic principle is to axomitize the system to be verified using a mathematic notation. Then, properties are theorems to be demonstrated from these axioms, possibly by means of intermediary intermediate lemmas or theorems. A "proof assistant" is used to help engineers to explore the "demonstration space". However, this assistant cannot perform the proof automatically.

The theorem prooving approach allows us to demonstrate that the system respects the desired properties. In addition, we know the validity conditions for these proofs, such as the parameters of the initial state, required for verifying the properties. They, therefore, are extremely useful data for the system designers.

However, this technique has several drawbacks. The first one is that it is difficult to put into practice. This approach requires highly qualified engineers mastering both the considered application domain and the demonstration techniques. Another difficulty refers to the absence of a diagnosis demonstration that cannot be performed (i.e. the exploration carried out by the proof assistant fails). Only an extremely high degree of expertise permits, in certain cases, to understand whether the absence of proof is due to the system itself or due to a modeling error.

1.5.2. *Formal verification by model-checking*

The *model-checking* principle [QUE 82, CLA 86, CLA 00] is very simple. The system is described using an executable formal notation, where a state is generally represented as a vector if values (e.g., the state of the variables of a process). This system is then simulated exhaustively, which can be done because the nature of the state allows us to tell, for each explored system configuration, whether it has been encountered before or not. Thus, if the system is finite, we may explore its state space exhaustively and look for the properties we wish to check (there are also model-checking techniques dedicated to infinite systems [DEM 11]).

The main advantage of this technique is that it is completely automatic; its use does not require the engineers to have any specific knowledge (only the specification language must be well known). Furthermore, in most cases, the response is either "yes, the property is verified", or "the property is not verified, but here is a counter-example that leads to the violation of the property". This information is a useful diagnosis, which can be used directly by an engineer on the basis of his knowledge of the system only.

Model-checking also suffers from a number of disadvantages. The first one is the combinatorial explosion of the number of states in complex systems [VAL 98]. This is particularly true when we introduce parallelism or when we wish to analyze time-based constraints. To avoid this problem, we must develop specific techniques that only function in certain cases. The engineer must then adapt their specification or use certain tools rather than others. Such cases temper the automatic use of model checking since, a deeper understanding of the underlying techniques is required to operate them.

The other problem is that the system cannot be verified in a parameterized way, as in the case of a proof, but only for given an initial configuration. This can make it difficult to identify the conditions that prompt a system to observe the desired properties.

Finally even if the counter example is small compared to the system complexity, the user may recover a trace of the 10^8 steps that are indeed difficult to analyze. Despite this, it is a good downsizing factor in comparison with a system that comprises, say, 10^{80} states.

In the rest of this section, we will look into the conditions that enable model-checking to verify systems.

1.5.3. *The languages to express specifications*

Expressing the specifications is a crucial problem because the verification algorithms function on the basis of the expression of the specification. In general, we can distinguish two parts in the system specification: the system itself and the properties it must respect. Let us note that, in general, the expression of the properties is harder than it appears.

1.5.3.1. *System modeling*

To be useful for the verification, we need a language with a formally defined operational semantic. Thus, the language must not only be executable but the notion of progression also needs to be formalized. These languages usually require two elements: the notion of state and the notion of transition between two states. More particularly, in order to apply certain algorithms, we need the transition between two states to be reversible (which is not always the case in the "real world").

Thus, it is difficult to directly apply model-checking to a programming language whose semantic is too sophisticated (let us note that there are works aiming at directly verifying the programs, however this will be tackled later). For example, memory allocation is typically a complex problem that can be addressed using model-checking. *A priori*, automata are the basic language for model-checking, given that the state space is nothing other than an automaton whose nodes are the configurations (the states of the system – the vector we mentioned above) and the transitions and the relations between these states.

Thus, we obtain the state space of a system from a product between the automata that represent its components. However, given that automata are not necessarily the easiest model to use, it is often necessary to use "automata generators", such as Petri nets [DIA 09], or languages with possibilities restricted to a well-defined behavioral model such as PROMELA [HOL 97, HOL 04], CSP [HOA 85] and FIACRE [BER 08].

Finally, recent studies concern the transformation of high-level languages into formal languages. Let us mention:

– the transformation of C [ZAK 08, JIA 09] programs or Java programs [VIS 05] (in general, a language subset) into PROMELA;

– the transformation of some UML diagrams into Petri nets [KOR 10];

– the transformation of AADL specifications into Petri nets [REN 09] or FIACRE [COR 10].

1.5.3.2. *The expression of properties*

Once we know how to express the state of a system, the properties can be specified as logical formulas expressing constraints on the components of the vector that describes a state. We thus obtain an atomic expression allowing us to characterize a state pattern such as the formula below, which implies three variables V_1, V_2 and V_3:

$$V_1 < 4 \ \wedge (V_2 > 5 \ \vee \ V_3 = 5)$$

This formula can be assessed throughout the exploration of the state space of the system. We call these safety or reachability properties because they aim at identifying a state that observes the given pattern. When such a state is reached, the algorithm stops the construction and looks for a path between the initial state and the state that is being characterized: this is the counter example. If the state space is explored without encountering the pattern, then the property is not verified. We can thus verify the absence of the undesired states in the system.

However, atomic expressions cannot express causal relationship between states. For example:

$$M_{received} = request \Rightarrow in\ the\ future, M_{sent} = response$$

This formula relates all of the states that correspond to the reception of a request to the fact that in the future, the server will necessarily send a response. These are temporal formulas (in the causal sense of the term). The atomic expressions are connected to this kind of formulas via temporal logic operators [WIK 12]. There are several classes of temporal logic, the most well known being the *linear temporal logic* (LTL) and *computation tree logic* (CTL).

For the management of time properties (i.e. involving time), engineers have developed TCTL or TLTL, which are extensions of CTL and LTL. CTL (and LTL) operators are then annotated with time intervals. Thus, the previous request, when timed, becomes, for example:

$$M_{received} = request \Rightarrow in\ less\ than\ 10\ time\ units, M_{sent} = response$$

The formula will not be verified unless the sending of the response follows the reception of the request in less than 10 time units.

A logic standardized by the ISO was elaborated in the 2000s: property specification language (PSL) [EIS 06, IEE 10]. It integrates notions coming from classic temporal logic and allows for time management or probability management.

There are much more than one algorithm required for assessing different types of formulas which varies in complexity. The reachability is the simplest (complexity in the size of the state space), then follow the algorithms for temporal logic formulas and, finally, those concerning timed temporal logic or probabilistic temporal logic. In certain cases (for instance, for timed systems), properties are undecidable: there is no algorithm that can systematically solve the problem.

The main difficulty in carrying out the model-checking approach lies in the capacity of the engineers to easily express the requirements that must be verified on a system. Temporal logic classes have a large power of expression and allow for a rigorous expression. However, in practice, they are difficult to handle in an industrial context because they demand great expertise from engineers. Indeed, a requirement can reference numerous events, which are connected to the execution of the model or of the environment, and it depends on an execution history that must be considered when at verification time.

One solution to this problem is the use of dedicated languages, which allow us to express properties and abstract certain details, at the cost of reducing the expressiveness. Numerous authors have made this observation and some of them [DWY 99, SMI 02, KON 05] have proposed to formulate properties with the help of definition patterns. A pattern is a textual syntactic structure that allows a mode of expression that is closer to the languages used by engineers.

Another way of simplifying the expression of the requirements comes from the fact that, in the requirement documents, they are often expressed in a given context of the system execution. The requirements are associated with specific phases of the system execution. In [KON 05], the authors have proposed to identify the scope of a property by enabling the user to specify the temporal context of the property with the help of the operators (*global, before, after, between, after-until*). These allow us to associate the requirements to a particular temporal context of the execution of the model to be validated. The scope indicates if the property must be considered, for example, during all of the execution of the model, before, after or between some occurrence of events. The analysis presented in Chapter 8 (Part 3) is inspired from this notion.

Let us note another technique adapted to temporal properties involving the detection of evens: the approach based on observation models [HAL 93]. The main idea is to overload the model with elements that are mere observers (non-intrusive), which observe some particular states of the system. It has been proved that with such observers, we can express these formulas in the form of an accessibility property [KUP 99]; therefore, a property can be verified by simpler algorithms. However, this needs a modification which is sometimes complex in terms of specification. We must also ensure that this has no side effect on the behavior of the system (i.e. that the observer must remain neutral and should not hamper certain behaviors).

This procedure has been popularized with the LUSTRE [HAL 91] language and has been reused by the UPPAAL tool [UPP 12]. The expression of an invariant property by means of an observer is simpler, yet its complexity is correlated to the complexity of the initial property (it can be up to 20 states for realistic observers).

1.5.4. *The actual limits of formal approaches*

The formal approaches are gradually becoming more present in several industrial sectors, becoming more and more indispensable for ensuring higher reliability. An indicator is the adoption, in the aeronautic standard DO178C, of formal techniques for system design. However, several questions yet remain open.

The first one considers the connection between the system and the model. Verification is carried out on a specification and not on the system in itself, often expressed in inappropriate terms or being too complex. Therefore, we must ensure consistency between the two, or else properties demonstrated on the model may not be true on the real system. This means that we need to use a very rigorous methodology.

To ensure this consistency between the system and the model, model engineering proposes approaches that involve transformations (several approaches are described in this book). The formal specification is thus generated from a high-level model that also serves to produce the code of the final application. This raises two issues:

– The approaches via transformations must preserve the execution semantic between the source model and the target model, which is difficult to demonstrate.

– The formal specification therefore become too complex: again we meet the problem of a combinatorial explosion, which is specific to model-checking approaches (or to their equivalent for proof-based approaches).

Finally, the use of formal methods requiring complex software tools raises the issue of their certification when they contribute to the elaboration of certified programs. One such successful experience has been carried out with the SCADE code generator, which is certified, and produces code that does not require any further certification efforts. The high licensing cost involved in certifying the code generator for SCADE, however, makes it rarely used. Specialists are debating about potentially using methodologies that allow for the use of uncertified tools in the development of certified software, an analysis of the validity of the results of these tools thus becoming indispensable. For a code generator, a simplified procedure for certifying generated programs, therefore, needs to be maintained.

These elements, along with the need for highly qualified engineers, are obstacles to the adoption of formal methods on a wider scale. It does not, however, stop progression of formal method use in computer science thanks to the outcome of research on the one hand and the increasing demand for system reliability, on the other hand.

1.6. Methodological aspects of the development of embedded computer systems

Since World War II, and then with the big space programs of the 1960s, the willingness to master the development of systems has grown considerably. To do this, we must have the various teams as well as the various specialties involved to collaborate more efficiently in the design and in the production of these complex systems. Thus, the actors concerned had to formalize the nature of the activities required to pass from more or less well defined stakeholders needs to a real system.

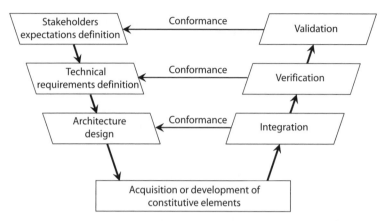

Figure 1.2. *V development cycle as proposed by the AFIS*

One of the most famous examples is the V development cycle (see Figure 1.2). This has become a standard in the industry since the 1980s. It distinguishes between:

– a downward specification/design branch, which groups together all the activities starting from the "need" to the "on paper" definition of the technical solution;

– an upward branch that covers the integration, verification and validation of the complete system;

– between the two, a horizontal passage corresponding to the acquisition or the development of constitutive elements (with, if necessary, other overlapped V cycles; the procedure being iterative).

This vision is of course simplified because it presents a perfectly sequential succession of activities. In reality, it also goes backward and iterations are inevitable. They can come from certain gaps or from evolutions of the requirements (certain aspects will only be emphasized via questions that arise during the design) as well as from interactions between specialities and/or from the search for better compromises, not to mention the problems detected during the integration or during tests.

This has led to more sophisticated development lifecycles (incremental, spiral, evolving, product lines, etc.), in particular for the software intensive systems. Nowadays, all of standards in engineering implement the following distinction between two dimensions:

– the *phases*, which punctuate the project, relying on an adapted development lifecycle model and facilitating the understanding of the entire process (for instance, the specification phase of the system, which results in supplying the technical specification);

– the *process* (set of connected activities), which indicates the activities that must be carried out; they are, essentially, independent of the type of system and the considered application domain.

The same type of activity may concern several phases. Thus, design activities may be necessary very early in the process, for instance for designing prototypes to illustrate and elaborate the requirements. Similarly, the verification and validation processes obviously apply to the final system, but they can also regard the results of other phases in order to detect problems or drifts as soon as possible (usually forgotten requirements/constraints or the impact of an undesirable event). The objective is to limit the impact of these problems on the production costs (money, time) of the system. It is the responsibility of the project management to select, organize and plan the activities that must be carried out.

1.6.1. *The main technical processes*

The main technical processes considered in engineering standards are, notwithstanding close variations in scope and terminology, based on the same pattern. Let us note that these processes indicate what needs to be done, but they do not suggest how. This is the objective of the methods, which are closely related to application domains as well as to specialties. In the processes usually considered, we find the following elements:

– The stakeholder requirement definition: this involves the collection, analysis and formalization, with the involved stakeholders (clients, users, support), of what the future system will have to do.

– The requirements analysis: this involves the translation of the stakeholders' requirements needs into measurable technical terms, without considering any possible implementation. The system is then seen as a "black box" whose interaction with its environment is analyzed. It is also in this stage that "non-functional" requirements (safety, security, reliability, availability, etc.) as well as any potential specific constraints (regulatory, environmental, etc.) are introduced.

– The functional design: this stage defines the functional architecture of the system (independent of any technological constraint), the allocation of requirements to functions as well as the description of their characteristics and behavior.

– The physical design: this stage partitions functions of the system (we speak of logical architecture), defines the physical architecture of the solution and allocates functions to the respective elements (hardware, software and, finally, human elements). We specify here the components to be developed along with their interfaces.

– The system integration: this involves assembling in one single system the set of components that had been developed and acquired separately.

– The system verification[2]: this stage concerns the verification actions to be done by the system designer/developer. They enable the system designer to ensure that his or her product is "well-made", with respect to development rules or standards, etc. (he or she has made the system right).

– The system validation: this involves actions that enable the client and the end user to ensure that the product responds correctly to the stakeholders' needs and requirements identified in the first stage (they have the right system).

1.6.2. *The importance of the models*

System development requires the collaboration of multiple actors and teams that are sometimes important and very varied specialties. We need, therefore, means of support, ensuring a smooth communication among all parties involved. This is the role of the models, which are abstract and partial representations of the system from one point of view and with a level of granularity that facilitates the study of some of its characteristics (properties, behavior, etc.). Each model is, therefore, designed with one precise objective in mind; however, no model is enough by itself to translate the complexity of a system.

There are several modeling languages, more or less specialized. This book deals with the languages adapted to the description and analysis of embedded systems, which are characterized by strong expectations in terms of response time, safety and security:

– SysML is a graphic modeling language used for capturing the structural, functional and behavioral aspects of the systems incorporating hardware, software, people, and procedures. It can be used in support of specification activities, analysis activities, design and verification-validation activities.

– UML/MARTE and AADL are modeling and analysis languages dedicated to real-time embedded systems. They allow us to consider non-functional properties (time constraints, safety constraints, security constraints, etc.) and to verify properties such as scheduling, good transmission of messages and the right sizing of the hardware. Their use takes place rather at the end of the design activities. They are neither methodology nor tools.

---

2 Do not mistake with "formal verification" in the sense of section 1.5.

Assigning a main role to models, model-driven engineering aims to guarantee the consistency of the elements manipulated by the different stakeholders throughout the system development lifecycle. The main idea is to gradually refine the models throughout the requirement definition and analysis, as well as design analysis and to rely on model transformation techniques in order to guarantee that a global consistency is maintained and their properties are preserved throughout each stage.

The semantic differences between the modeling languages specific to certain domains and the limited interoperability between the tools are still significant obstacles in the way of achieving this aim.

1.7. Conclusion

In this chapter, we have sketched an overview of the modeling and the analysis of embedded computer systems. Far from being exhaustive, this overview seeks to show, the richness of these activities: the variety of the description methods, the analysis techniques and the need to combine them via a rationalized process in connection with the respective constraints (i.e. domain constraints, normative and regulatory constraints).

This chapter enables the reader to better contextualize each of the notations presented. The book is devised around the same plan: the presentation of the notation, its use for representing a complex problem: a pacemaker, analysis and code generation. Reading these chapters will also enable the reader to understand when and how to use the different notations.

At first sight, we could think that SysML is useful in the early phases of the modeling process. Being a notation for system engineering, we will have to complement SysML with a notation that is closer to implementation considerations. We must therefore chose between UML/MARTE and AADL:

– UML/MARTE is an obvious candidate because it is derived from UML, as SysML is UML/MARTE which allows us to follow the modeling of the system by explicitizing some aspects of the system. However, it has various limitations connected with UML: a lack of a clear process, too much room for interpretation and a semantic that is too large, so that in certain cases we will have to use a subset of it.

– AADL, allows for a more precise modeling, with a semantic that is restricted to critical embedded systems. Furthermore, it is already connected to several analysis tools. However, since it is outside of the UML framework, we must build a more complex traceability between SysML and AADL.

We can thus imagine several combinations: UML/MARTE for allowing us to model the PIM and then the PSM of the system considered, then AADL in order to have a coherent view of all the elements, paving the way for the integration effort, as well as the verification and validation phases. It is better, if we limit ourselves to only one notation, UML/MARTE or AADL, to model the system and perform analysis and/or code generation.

As we have seen, choosing a notation is not a simple matter, and it is, above all, a choice that must be dictated by the problem we need to solve, by the tools that are available as well as by how familiar the engineer is with the notation.

This is the overarching goal of this book, enabling you, the reader, to make such a choice, by analyzing the same case study using these three notations.

1.8. Bibliography

[ABR 96] ABRIAL J.-R., *The B Book – Assigning Programs to Meanings*, Cambridge University Press, 1996.

[AFF 08] AFFELDT R., KOBAYASHI N., "A Coq library for verification of concurrent programs", *Electronic Notes in Theoretical Computer Science (ENTCS)*, vol. 199, pp. 17–32, 2008.

[BER 08] BERTHOMIEU B., BODEVEIX J.-P., CHAUDET C., *et al.*, "Verifying dynamic properties of industrial critical systems using TOPCASED/FIACRE", *ERCIM NEWS, Special Issue on Safety-Critical Software*, September 2008.

[CLA 86] CLARKE E., EMERSON E., SISTLA A., "Automatic verification of finite-state concurrent systems using temporal logic specifications", *ACM Transactions on Programming Languages and Systems*, vol. 8, no. 2, pp. 244–263, 1986.

[CLA 00] CLARKE E., GRUMBERG O., PELED D., *Model Checking*, MIT Press, MA, 2000.

[COR 10] CORREA T., BECKER L.B., FARINES J.-M., *et al.*, "Supporting the design of safety critical systems using AADL", *15th IEEE International Conference on Engineering of Complex Computer Systems (ICECCS)*, IEEE Computer Society, pp. 331–336, 2010.

[DEM 11] DEMRI S., POITRENAUD D., "Verification of infinite-state systems", HADDAD S., KORDON F., PAUTET L., PETRUCCI L., (eds), *Models and Analysis in Distributed Systems*, ISTE Ltd, London and John Wiley & Sons, New York, pp. 221–269, 2011.

[DIA 09] DIAZ M., (ed.), *Petri Nets, Fundamental Models, Verification and Applications*, ISTE Ltd, London and John Wiley & Sons, New York, 2009.

[DWY 99] DWYER M.B., AVRUNIN G.S., CORBETT J.C., "Patterns in property specifications for finite-state verification", *21st International Conference on Software Engineering*, IEEE Computer Society Press, pp. 411–420, 1999.

[EIS 06] EISNER C., FISMAN D., *A Practical Introduction to PSL* (Integrated Circuits and Systems), Springer, 2006.

[FIM 74] FIMMEL R.O., SWINDELL W., BURGESS E., SP-349/396 PIONEER ODYSSEY, 1974. Available at history.nasa.gov/SP-349/contents.htm.

[HAL 91] HALBWACHS N., CASPI P., RAYMOND P. *et al.*, "The synchronous dataflow programming language LUSTRE", *Proceedings of the IEEE*, pp. 1305–1320, 1991.

[HAL 93] HALBWACHS N., LAGNIER F., RAYMOND P., "Synchronous observers and the verification of reactive systems", in NIVAT M., RATTRAY C., RUS T., SCOLLO G. (eds), *Third International Conference on Algebraic Methodology and Software Technology, AMAST'93*, Twente, Workshops in Computing, Springer-Verlag, June 1993.

[HOA 85] HOARE C., *Communicating Sequential Processes*, Prentice Hall, 1985.

[HOL 97] HOLZMANN G., "The model checker SPIN", *Software Engineering*, vol. 23, no. 5, pp. 279–295, 1997.

[HOL 04] HOLZMANN G., "An overview of PROMELA", *The SPIN Model Checker*, Addison-Wesley, pp. 33–72, 2004.

[IEE 10] IEEE 1850, IEEE Standard for Property Specification Language (PSL), 2010.

[ISO 02] ISO/IEC 13568,   Z formal specification notation – syntax, type system and semantics, 2002.

[JIA 09] JIANG K., JONSSON B., "Using SPIN to model check concurrent algorithms, using a translation from C to Promela", *2nd Swedish Workshop on Multi-Core Computing*, Uppsala, 2009.

[KON 05] KONRAD S., CHENG B., "Real-time specification patterns", *27th International Conference on Software Engineering (ICSE05)*, St Louis, MO, 2005.

[KOR 10] KORDON F., THIERRY-MIEG Y., "Experiences in model driven verification of behavior with UML", *Foundations of Computer Software, Future Trends and Techniques for Development, 15th Monterey Workshop 2008*, Budapest, Revised Selected Papers, Lecture Notes in Computer Science, vol. 6028, Springer, pp. 181–200, 2010.

[KUP 99] KUPFERMAN O., VARDI M., "Model checking of safety properties", *International Conference on Computer Aided Verification*, Lecture Notes in Computer Science, vol. 1633, Springer, pp. 685–685, 1999.

[OMG 12a] OMG, The Official OMG SysML site, 2012. Available at www.omgsysml.org.

[OMG 12b] OMG, UML profile for MARTE modeling and analysis of real-time embedded systems, 2012. Available at www.omg.org/spec/MARTE.

[QUE 82] QUEILLE J.-P., SIFAKIS J., "Specification and verification of concurrent systems in CESAR", *Proceedings of the 5th Colloquium on International Symposium on Programming*, Springer-Verlag, London, UK, pp. 337–351, 1982.

[REN 09] RENAULT X., KORDON F., HUGUES J., "Adapting models to model checkers, a case study: analysing AADL using time or colored petri nets", *Proceedings of the 20th International Symposium on Rapid System Prototyping*, IEEE Computer Society, Paris, pp. 26–33, June 2009.

[SAE 09] SAE, Architecture analysis & design language V2 (AS5506A), 2009. Available at www.sae.org.

[SMI 02] SMITH R., AVRUNIN G., CLARKE L. *et al.*, "Propel: an approach supporting property elucidation", *24th International Conference on Software Engineering (ICSE'02)*, ACM Press, St Louis, MO, pp. 11–21, 2002.

[UPP 12] UPPSALA AND AALBORG UNIVERSITIES, UPPAAL Home, 2012. Available at www.uppaal.org.

[VAL 98] VALMARI A., "The state explosion problem", REISIG W., ROZENBERG G. (eds), *Lectures on Petri Nets 1: Basic Models*, Lecture Notes in Computer Science, vol. 1491, Springer-Verlag, pp. 429–528, 1998.

[VIS 05] VISSER W., MEHLITZ P.C., "Model checking programs with Java Pathfinder", *Model Checking Software, 12th International SPIN Workshop*, Lecture Notes in Computer Science, vol. 3639, Springer, p. 27, 2005.

[WIK 12] WIKIPEDIA, Temporal Logic, 2012. Available at en.wikipedia.org/wiki/Temporal_logic.

[ZAK 08] ZAKS A., JOSHI R., "Verifying multi-threaded C programs with SPIN", *15th International SPIN Workshop*, Lecture Notes in Computer Science, vol. 5156, Springer, pp. 325–342, 2008.

Chapter 2

Case Study: Pacemaker

2.1. Introduction

The aim of this chapter is to present the case study that will serve as a common example to illustrate the different modeling techniques presented in this book. We have taken a pacemaker whose specifications have been published by Boston Scientific [BOS 07]. This case study is part of a great challenge, whose goal is to stimulate research around formal methods and associated modeling techniques, through a series of challenges to the academic and industrial communities [WOO 06].

This case study has been treated in many ways; we can refer to modeling efforts done using Z [GOM 09], Event-B [MÉR 09], VDM [MAC 08] and process algebra [TUA 10]. Finally, the winning team in the competition published the report detailing their approach in 2009, combining theorem proofs (including PVS [MAN 09]) and SAT solver, available on the SCORE website[1]. Parallel to this, an electronic implementation was developed by the University of Minnesota [NIX 09].

We have chosen this case study since it provides a concrete example of a well-known embedded system, covering a wide range of concerns regarding procedural aspects (installation, pacemaker initiation), interactions between the operator/patient/device and its behavior.

Chapter written by Fabrice Kordon, Jérôme Hugues, Agusti Canals and Alain Dohet.
1 score-contest.org/projects.php#Pacemaker

2.2. The heart and the pacemaker

In this section, we give a brief presentation of the system studied: a heart stimulator or pacemaker. A pacemaker is a small electronic system intended to assist the heart in maintaining a regular rhythm. The pacemaker is surgically implanted in the patient's torso. Leads are positioned on the heart muscles. The pacemaker itself is positioned under the skin, close to the shoulder. The pacemaker operates in the immediate vicinity of the heart.

First, we present the function of the heart before returning to the elements of a pacemaker.

2.2.1. *The heart*

The role of the heart is to enable the blood to flow to the different organs in the human body. It acts like a pump that functions without interruption throughout one's life. The function of the heart can be reduced to that of a mechanical and electrical system. Its behavior is regulated as a function of effort. Hence, a control "algorithm" exists.

At rest (i.e. in the absence of physical effort), the heart rate is approximately 68 beats per minute for a flow rate of 4.5–5 L of blood per minute. During a lifetime, the heart can beat more than two billion times.

The heart's mechanical system (the pump) is stimulated electrically to trigger muscle action. The heart (see Figure 2.1) comprises four compartments: the left and right ventricles and atria (or lobes). These compartments contract and release periodically, controlled by electric impulses. The atria form one unit and the ventricles form second unit.

Each of its movements involves a sequence of events that are collectively known as the "cardiac cycle". This consists of three major stages: auricular systole, ventricular systole and the diastole.

– During the auricular systole, the atria contract and push blood into the ventricle (active filling). Once the blood has been expelled from the atria, the auricoventricular valves between the atria and the ventricles close. Blood continues to flow into the atria. This avoids a reflux of blood toward the atria. The closing of the valves produces the familiar sound of the "heartbeat".

– The ventricular systole involves the contraction of the ventricles, pumping blood into the circulatory system. In fact, very briefly in the first moment, the semilunar valves are closed. Once the pressure inside the ventricles exceeds the arterial pressure, the semilunar valves open. Once the blood has been pushed out, the two semilunar valves – the pulmonary valve on the right and the aortic valve on the left – close. In this way, blood does not flow back into the ventricles. The closure of the semilunar valves produces a second cardiac sound, which is sharper than the first and then the blood pressure increases.

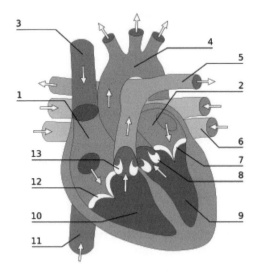

1) Right atrium

2) Left atrium

3) Superior vena cava

4) Aorta

5) Pulmonary artery

6) Pulmonary vein

7) Mitral valve (atrio ventricular)

8) Aortic valve

9) Left ventricle

10) Right ventricle

11) Inferior vena cava

12) Tricuspid valve (atrio ventricular)

13) Semilunar valve (pulmonary)

Figure 2.1. *Anatomical sketch of the heart (Wikipedia©)*

– The diastole is the relaxation of all parts of the heart, which allows the (passive) filling of the ventricles (more than 80% filling, in usual conditions), via the left and right atria from the vena cava and pulmonary vein. The atria fill gently and blood drains into the ventricles.

In normal functioning, the human body has control mechanisms for heart stimulation in the central nervous system and the endocrine system. These two systems play the role of a "natural" pacemaker. Two nodes of the nervous system play a key role in electric stimulation: the sinoatrial node and the atrioventricular (AV) node. The sinoatrial node receives an initial electrical impulse, about 1 mV, which is then transmitted to the AV node that amplifies it, stimulating the ventricles.

A pacemaker can be implanted in a patient whose own body cannot by itself ensure a regular heart function that matches the patient's physical exertion. In this case, the pacemaker substitutes for the normal instructions transmitted by the body. We present its functioning in the following section.

2.2.2. *Presentation of a pacemaker*

A pacemaker is an electronic device implanted in the heart of a patient suffering from bradycardia (from the Greek *bradus* = slow and *kardia* = heart). Bradycardia is a condition characterized by a much slower heart rate than normal, in general one that is less than 60 beats per minute, or the opposite of tachycardia, which is characterized by a too high heart rate, above 100 beats per minute in the absence of significant physical effort.

The pacemaker's purpose is to emit electrical signals directly to the heart. Since the heart is a muscular structure, receiving correctly measured signals will correct its movement.

The basic elements of a pacemaker are:

– Leads: one or more covered metal wires (most often two) that transmit electrical signals between the heart and the pacemaker. Each probe can have one or two points of contact with the heart, and is therefore known as a uni- or bipolar leads.

– Pulse generator (PG): the pacemaker is both the power source and also the controller and monitor of heartbeats. It therefore contains an accumulator and a unit that generates a signal and analyzes the heartbeat, and also emits an electrical signal. It is implanted near the heart.

– Device controller monitor (DCM): this is an external unit that interacts with the pacemaker via a wireless connection. It contains the pacemaker's application part, which makes the decision whether to emit a signal as a function of the patient's heartbeat.

– Accelerometer: it is located in the pacemaker and measures acceleration and body movements of the patient in order to adjust the stimulations.

For a single-electrode pacemaker, the electrode is attached to the atrium of the right ventricle. The pacemaker has several operational modes to control the movement of the heart, as detailed in the specification document [BOS 07]. These parameters are related to real-time stresses and causality (action/reaction) stresses, which allows the cardiac rhythm to be controlled.

This quick presentation of the pacemaker does not cover its different modes of use, or the multiple interactions occurring between the patient, the pacemaker and the medical team. In particular, we have not presented the process stages of implantation, diagnosis or their associated errors. We recommend the reader refers to the book by Barold *et al.* [BAR 10] for a more detailed account.

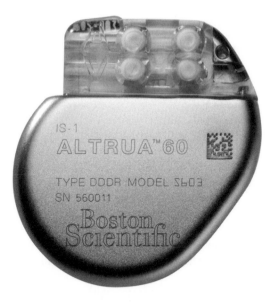

Figure 2.2. *Boston Scientific pacemaker*

2.3. Case study specification

In this section, we are interested in the description of the pacemaker in the [BOS 07] specification, and will not look at the graphical interface presented in section 4 of the document.

This section also aims to introduce the system requirements, which will be explored later through the different modeling languages discussed in this book. The requirements are numbered, and labeled as follows:

(1)/S (*Client aim*) The pacemaker functions correctly.

An identifier in the margin "(XX)/Y" indicates that the paragraph concerned is a requirement, defined in natural language. Italic text at the start of the paragraph defines the category of the requirement: the pacemaker functions correctly. The suffix indicates the level of the requirement, "S" for system and "P" for the pacemaker itself.

We need to refine and clarify what "correctly" means. This is addressed in the following sections.

We present here an excerpt of the requirements as they are defined by Boston Scientific. An attentive reader will note that some of these requirements are in fact definitions, which we have kept. These requirements will be sorted in the design phase.

2.3.1. *System definition*

The pacemaker meets the following patient requirements:

– Implantation.

– Ambulatory phase, when the system is calibrated.

– Follow-up with the patient.

– Explantation.

The system provides stimulation at a rhythm that is adapted for the patient, on two of the heart chambers, data on its functioning in a log and diagnostics on its proper functioning.

Its analysis functions allow the following measurements: probe impedance, pulse threshold, measurements of the heart, battery status, pulse emission and accelerometer calibration.

Since the pacemaker is a medical system, its visible components are defined and standardized, which makes it straightforward to swap patients and medical teams.

Therefore, the pacemaker comprises three main components: the actual pacemaker, that is the PG; the DCM and its associated control software; and the leads. The cardiac magnet in the shape of a *donut* is a minor component. Note that through a misuse of language, pacemaker most often refers only to the PG.

The PG controls and regulates the patient's heartbeat, thereby providing a remedy for bradycardia. The PG provides a programmable means by which one of two heart chambers can be stimulated, adapted to the patient's rhythm, whether this be permanent or temporary.

In an adaptive fashion, an accelerometer is used to measure the patient's physical activity. This measurement is used to calculate the stimulus to emit.

The PG is programmed and interrogated via telemetry by the controller (DCM). This enables the doctor to change the operational mode or certain parameters in a non-intrusive way after implantation.

The PG provides the following data: sensor output data and histograms for the *atria* and ventricles.

The PG, in conjunction with the DCM, provides the following diagnostic information: real-time telemetry, electrograms (EGM), impedance of the heart to the leads and battery status.

The DCM is the primary support for the treatment; it is used by different medical teams (implantation, ambulatory and follow-up). It communicates via a communication protocol and a dedicated hardware layer. The DCM is composed of a hardware platform and a programmable software block.

The DCM provides the following functionality: programming and interrogating the pacemaker, sending an immediate stimulation command, storing historical sensor data and displaying historical and calibration data.

In addition, it has auxiliary functions: function indicators using LEDs, collecting data for electrocardiograms, monitoring the battery status, sending output to a display device, etc. These functions are not considered here.

The leads connected to the patient's heart enable the heart's electrical activity to be detected and stimulated. The leads are connected to the PG via an adapted connector. All IS-1 bipolar leads are supported.

The pacemaker and the leads are implanted by a hospital medical team. Follow-up with patients is done by nurses and technicians, under the supervision of a doctor.

The pacemaker should meet the electrical safety standard requirements. It must work in all stages of its lifecycle, and must withstand sources of interference such as fluoroscopy, control machines for anesthesia, body fluids and defibrillators.

2.3.2. System lifecycle

(2)/S   The lifecycle of the pacemaker is as follows:

– Pre-implant: the system is manufactured following good practices as required by certification authorities and health regulations. During this stage, the nominal parameters of the PG are configured.

– Implant: the pacemaker is placed inside the patient's body. During this stage, the DCM unit is used to study the system, check the batteries, test and configure the pacemaker, program the system and evaluate the different parameters read by the leads. The implant procedure involves the following stages: checking all the equipment, implanting the leads, evaluating the signal levels from the leads, programming the system before implantation, putting the leads into the body, connecting the leads and testing them and implanting the PG.

– Initial follow-up: during this stage, a series of tests can be performed: studying the system to obtain its parameters, reprogramming and printing out an overview of the state of the system for storing the patient's records.

– Routine follow-up: the programmable system goes through the following actions, known as routine: monitoring the system state, checking the battery state and bradycardia parameters of the patient, measuring the P and R waves, testing the stimulation threshold and probe impedance, reviewing the historical data and associated histograms and deleting them, and printing out the test reports. If the PG parameters are modified, these can be checked before implantation:

– Ambulatory: the stimulation functions and measurements are available.

– Explantation: once the PG is retired, it is cleaned and returned to its manufacturer. A fault analysis is performed. The state of the PG, when it is retired, depends on the faults that it has experienced, and/or on the battery status.

– Destruction: the pacemaker is sent back to its manufacturer. It must not be incinerated since some of its components may explode at high temperatures.

2.3.3. System requirements

In the following, we consider the pacemaker system taken alone, in the case of treatment of bradycardia.

(3)/S   The pacemaker must be clearly identified by a model number, its description, the supported functions and the connectors used.

(4)/S   The pacemaker has a manual for its functioning and software configuration, according to its measurements. We do not present this part.

(5)/S   The pacemaker and the DCM must communicate with the PG. Possibilities include magnetic induction, ultrasound, or by radio transmission within the legal limits.

(6)/S   The bipolar leads on the atrium and ventricle must be usable. Probe impedance must be between 100 and $2,500\,\Omega$.

(7)/S   Pulses generated by the pacemaker must be programmable in both amplitude and width.

(8)/S   Measurement of heart activity must be done using electrodes. The decision to detect a heartbeat must be based on measurements of the cycle length. The measurement must be done through a moving window.

(9)/S   The sensitivity of the leads must be a parameter that the doctor can adjust. Each probe must be independently programmable.

The pacemaker is a programmable electronic device. Its general behavior does not vary, and is defined by the intrinsic heart function. On the other hand, some situations are adjustable as a function of the patient's pathology. The types of bradycardia govern the different situations for which the pacemaker can be programmed.

(10)/S   (Naming) A pacemaker's operational mode is defined by a universal code composed of three or four characters. This code provides a clear indication of the active functions at any given moment. This sequence is the "pacemaker operational mode". Each chamber is given a letter: "O": nothing, "A": atrium, "V": ventricle, "D": equivalent to "A" + "V".

In this code, the first letter represents the chamber that is stimulated, the second is the chamber that is measured, the third indicates if a measurement is to be responded to and the last, an optional letter indicates whether the cardiac rhythm needs to be changed in response to physical activity, as measured by the accelerometer. The letter "X" is used to indicate all the other letters (i.e. "O", "V" or "D"). The letter "T" (*triggered*) indicates the delivery of a pulse, an electrical signal, while "I" (*inhibited*) indicates its absence after having measured the intrinsic activity of one of the heart's chambers.

The following requirements define the short-term behavior of the pacemaker, that is to say its behavior in a time period of the order of a few heartbeats.

(11)/S   (*No response to sensing (O)*) Stimulation without sensing, or asynchronous stimulation, corresponds to stimulation delivered without reference to the sensed measurements.

(12)/S   (*Triggered response (T)*) In this mode, a chamber is immediately triggered in response to a sensed measurement of the same chamber.

(13)/S   (*Inhibited response (I)*) In this mode, a sensed measurement of a chamber inhibits a pending stimulation of that same chamber.

(14)/S   (*Tracked response (D)*) In this mode, a sensed measurement triggers stimulation of the ventricle after a programmed delay, unless a movement of this same ventricle is detected during the interval.

The pacemaker defines a series of modes of greater maturity, defining a current state for the patient: permanent, temporary, immediate stimulation, magnet test and return to the initial state (*Power-On Reset* (POR)). These modes are mutually exclusive.

(15)/S   (*Permanent state*) This state is the normal functioning mode of the pacemaker. The configuration parameters for the patient are used.

(16)/S   (*Temporary state of bradycardia*) This state corresponds to a combination of parameters, which allows testing of the pacemaker parameters and provides diagnostic information on the patient. This mode is ended by one of the following actions: immediate stimulation or a command from the DCM unit.

(17)/S   (*Immediate stimulation*) An emergency stimulation for bradycardia called *Pace-Now* must be available. The parameters are as follows:

– The stimulation mode must be "VVI".

– The lower limit of stimulation must be 65 bpm ± 8 ms.

– The stimulation amplitude must be 5 V ±0.5 V.

– The stimulation length is 1 ms ± 0.02 ms.

– The ventricular refractory period is 320 ms ±8 ms.

– Ventricular sensitivity is 1.5 mV.

– The first stimulation must be emitted as soon as possible after two cycles plus 500 ms from the last activation.

– Once activated, this mode must be kept until a counter-order is sent by the DCM.

(18)/S   (*Reset to the initial state*) This operation is done when the battery level is too low to maintain effective behavior in the system. All the functions must be deactivated until the battery achieves a sufficient threshold level again. Once this threshold has been achieved, the pacemaker parameters are as follows:

– The stimulation mode must be "VVI".

– The lower limit of stimulation must be 65 bpm ±8 ms.

– The stimulation amplitude must be 5 V ±0.5 V.

– The stimulation length must be 0.5 ms ±0.02 ms.

– The ventricular refractory period is 320 ms ±8 ms.

– The ventricular sensitivity is 1.5 mV.

(19)/S   (*Magnet Test*) This mode is used to determine the battery status. A standard magnet, in the shape of a *donut*, must be detected by the pacemaker at a distance of 2.5 cm from the pacemaker's surface.

When the magnet is in place:

– The device should asynchronously stimulate at a fixed rate. The pacemaker mode should be "AOO" if the previous mode was "AXXX", "VOO" if the previous mode was "VXXX", "DOO" if the previous mode was "DXXX" or "OOO" if the previous mode was "OXO".

– At pacemaker initiation, the rhythm should be 100 beats per second; when the battery level drops to a certain level, the rhythm drops to 90 bpm, then 85 bpm and continues to fall as a measure of the battery voltage.

– When the magnet is removed, the pacemaker must assume that it is performing a pretest.

– We must be able to deactivate this mode, to avoid detecting the magnet.

(20)/S   (*Implantation data*)   The pacemaker must store the following information in memory:

– Pacemaker model, serial number and implantation date.

– Probe implantation date and probe polarity.

– Stimulation threshold and signal level.

– Probe impedance.

– The parameters for signals specific to the patient.

2.3.4. *Pacemaker behavior*

Section 2.3.3 presented the requirements for the pacemaker, its different functioning modes and its interaction with the patient and medical team. We presented the different pacemaker components and their associated parameters. In this section, we return to the pacemaker, studying its real-time behavior in the case of bradycardia treatment.

Treatment for bradycardia works by stimulating the patient's heart. This treatment must be adapted to the patient. There are therefore many parameters that are used as a function of the stimulation mode and the patient's needs.

First, we define several terms that are useful in characterizing the pacemaker behavior in response to a cardiac stimulation, or to its absence over a certain time.

(21)/P   (*Lower rate limit*) The lower rate limit (LRL) is the number of stimulations produced per minute (in either the atrial or ventricular cases) in the absence of intrinsic heart activity or controlled stimulation by the sensor at a higher rate.

The LRL definition varies with the context:

– When hysteresis is disabled, the LRL is defined by the longest interval allowed between two stimulations.

– In DXX or VXX modes, the LRL starts when a ventricular pulsation is detected, or when a stimulation is detected.

– In AXX modes, the LRL starts when an atrial pulsation or stimulation is detected.

(22)/P   (*Upper rate limit*) The upper rate limit (URL) is the maximum number of sensed events measured at the level of the atria – or pacing rate. The URL is the minimum interval time between a ventricular event and the next stimulation of the ventricle.

(23)/P   (*Atrio ventricular delay*) The AV delay is a programmable parameter that represents the interval between an atrial event (intrinsic or stimulated) and a ventricular pulsation.

In atrial tracking modes, ventricular stimulation must occur in the absence of a detected ventricular event at the latest after this delay period when the atrial pulsation rate is between the LRL and the URL. The delay can be fixed (absolute time) or relative.

The delay has a different value according to the triggering event:

(24)/P   (*Stimulated atrioventricular delay*) The stimulated delay should occur when the AV delay is initiated by an atrial stimulation.

(25)/P   (*Detected atrioventricular delay*) The detected delay should occur when the AV delay is initiated by an atrial pulsation.

(26)/P   (*Dynamic atrioventricular delay*) In the case where the delay is dynamic, its value is determined by each new cardiac cycle, as a function of the preceding cycles. This delay is obtained by multiplying the length of the preceding cycle by a constant stored in memory. Its value is limited by minimum and maximum values.

(27)/P   (*Calibrating the delay*) The parameter that is configurable through calibration allows reduction of the delay after detecting a following pulsation at the level of the atria.

(28)/P   (*Atrial Refractory period*) To limit detection errors, a refractory period, during which probe measurements are ignored, can be used.

Different refractory periods can be programmed, after a ventricular event (VRP), atrial event (ARP) and post-AV event (PVARP). In the last case, the refractory period is the time interval following a ventricular event when an event at the atrial level neither inhibits a spontaneous stimulation of the atria, nor triggers a ventricular stimulation.

The extended PVARP period is defined as follows: if this parameter is enabled, any premature ventricular contraction (PVC) event forces a refractory period for the

PG for a time period equal to this parameter; the value for the refractory period must return to normal by the following cardiac cycle.

(29)/P (*Noise response*) In the case of continuous noise, the pacemaker should be in an asynchronous stimulation mode.

(30)/P (*Atrial tachycardia response*) The atrial tachycardia response (ATR) aims to prevent tracking sustained, over-fast atrial contractions. In this mode, the PG must detect a tachycardia situation when the intrinsic rhythm of the atria exceeds the URL value for a certain length of time.

(31)/P (*Tachycardia detection*) Tachycardia detection functions as follows:

– The start of a tachycardia situation is detected when the cardiac rhythm is on average higher than the URL.

– The end of a tachycardia situation is detected when the cardiac rhythm is on average less than the LRL.

– The detection period should be short enough so that the treatment can be rapidly put in place.

– The detection period should be long enough so that the treatment is not triggered too often, or ineffectually.

(32)/P (*ATR Duration*) When a tachycardia is detected, the ATR algorithm enters the "ATR Duration" state. The PG must wait for a programmed number of cardiac cycles before entering the "Fallback" state. The algorithm exits this state as soon as the delay ends, or when the tachycardia ends.

(33)/P (*ATR Fallback*) If the tachycardia continues after the preceding delay: (1) the PG enters the "Fallback" state and thus enters VVIR mode; (2) the pulse rhythm is reduced to the LRL; (3) this mode ends as soon as the tachycardia situation ends and (4) the mode change is synchronized with a ventricular pulse or a detected event.

(34)/P (*Adapting the stimulation rate*) The pacemaker must be able to adjust the cardiac cycles as a function of the metabolic needs such as those measured by the accelerometer. The following parameters are defined:

(35)/P (*Maximum sensor rate (or MSR)*) The maximum sensor rate corresponds to the maximum stimulation rate in response to the sensor measurements: (1) this parameter is required in rate-adaptative modes and (2) is independently programmable from the URL.

(36)/P (*Activity threshold*) This threshold corresponds to the minimum activity detected by the accelerometer before the cardiac stimulation is altered.

(37)/P   (*Response factor*) The accelerometer should determine the new stimulation rate as a function of the patient's state of activity. The higher the value measured, the faster the response and modification of the stimulation rate.

(38)/P   (*Reaction time*) The accelerometer must determine the rate of increase in the stimulation rhythm. The reaction time corresponds to the time required to change from the LRL to the MSR.

(39)/P   (*Recovery time*) The accelerometer must determine the recovery rate that corresponds to the time interval required to change from the MSR value to the LRL, until activity drops below the activity threshold.

(40)/P   (*Hysteresis stimulation*) When this feature is activated, the generator pulses are delayed in order to allow a spontaneous stimulation of the heart. This mode is active when the stimulation mode is inhibited or followed, the cardiac rhythm is higher than the hysteresis rate limit; when the pacemaker is in AAI mode, a sensed event in the atria should activate hysteresis; or finally in the inhibited and followed mode with ventricle stimulation, an event is detected in the ventricle.

(41)/P   (*Rhythm smoothing*) Variations in the stimulation rhythm are smoothed so that they are not too abrupt. These parameters are set independently to increase and decrease the rate.

2.4. Conclusion

In this chapter, we have presented the common case study, which will be used in what follows in using each of the three formalisms chosen in this book: SysML, MARTE and AADL.

We have chosen the pacemaker for the case study, an electronic system that has the role of stimulating the heart of a patient suffering from bradycardia. We have detailed the pacemaker's structure and functioning through presenting the general architecture of the patients and the requirements that it must meet, as written by the experts.

This chapter cannot be a complete reference on the functioning of this device. The aim of this chapter is, above all, to provide a basic vocabulary that will be of use in the following chapters. Nevertheless, we have provided a summary of the requirements document, thereby illustrating the difficulty associated with embedded systems engineering: the communication of industry concepts to non-specialists, whose skills lie in software engineering.

In the following chapters, the authors show how these specifications are used in each of the chosen formalisms, particularly their use for the modes and behavior of the pacemaker, and also its use.

2.5. Bibliography

[BAR 10] BAROLD S., STROOBANDT R., SINNAEVE A., *Cardiac Pacemakers and Resynchronization Step by Step: An Illustrated Guide*, John Wiley & Sons, NJ, 2010.

[BOS 07] BOSTON SCIENTIFIC, "PACEMAKER System Specification", January 2007.

[GOM 09] GOMES A.O., OLIVEIRA M.V., "Formal specification of a cardiac pacing system", *Proceedings of the 2nd World Congress on Formal Methods*, FM '09, Springer-Verlag, Berlin, Heidelberg, pp. 692–707, 2009.

[MAC 08] MACEDO H., LARSEN P., FITZGERALD J., "Incremental development of a distributed real-time model of a cardiac pacing system using VDM", in CUELLAR J., MAIBAUM T., SERE K. (eds), *FM 2008: Formal Methods*, vol. 5014 of Lecture Notes in Computer Science, Springer, Berlin, Heidelberg, pp. 181–197, 2008.

[MAN 09] MANNA V.P.L., BONANNO A.T., MOTTA A., A simple pacemaker implementation, Report, 2009.

[MÉR 09] MÉRY D., SINGH N.K., Pacemaker's functional behaviors in event-B, Research Report, 2009.

[NIX 09] NIXON C., ULRICH T., LARSON C., *et al.*, "Academic dual chamber pacemaker", Report, 2009.

[TUA 10] TUAN L.A., ZHENG M.C., THO Q.T., "Modeling and verification of safety critical systems: a case study on pacemaker", *Proceedings of the 2010 Fourth International Conference on Secure Software Integration and Reliability Improvement*, SSIRI '10, IEEE Computer Society, Washington, DC, pp. 23–32, 2010.

[WOO 06] WOODCOCK J., "First steps in the verified software grand challenge", *Computer*, IEEE Computer Society, vol. 39, pp. 57–64, 2006.

PART 2

SysML

Chapter 3

Presentation of SysML Concepts

3.1. Introduction

For many years, system engineers have used modeling techniques. The best known of these are Structured Analysis and Design Technique (SADT) and Structured Analysis/Real Time (SA/RT), which date back to the 1980s, as well as many approaches based on Petri nets or on finite state machines. However, these techniques are limited in scope and expressiveness as well as difficult to integrate with other languages.

The development of Unified Modeling Language (UML) in the software arena, and efforts in industry to develop the tools to go with it, naturally led to considering the use of UML in systems engineering (SE). Because it was designed for object-oriented programming, early versions of UML were not very well suited to the modeling of complex systems, and therefore not used to support SE.

UML version 2 [OMG 05], released in 2005, introduced several new useful concepts and diagrams for systems modeling. In particular, the composite structure diagram, with the concepts of structured classes, parts, ports and connectors, makes it possible to describe the internal, structural interconnections of a complex system. The progress in sequence diagrams also makes it possible to describe interaction scenarios while progressively adding more levels of architecture. However, the psychological barrier of software-oriented language still exists: class, object, inheritance, etc.

Chapter written by Jean-Michel BRUEL and Pascal ROQUES.

The SE community wanted to define a common language for modeling, which is adapted to their particular issue, in the way that UML has become a common language for computer scientists. This new language, called SysML, takes much of its inspiration from UML version 2; however, it includes the possibility of representing system requirements, non-software components (mechanics, hydraulics, sensors, etc.), physical equations, continuous flows (matter, energy, etc.) and allocations (see section 3.7.2).

Version 1.0 of the SysML modeling language was officially adopted by the OMG on September 19, 2007. Since then, three minor revisions have been published: SysML 1.1 in December 2008, SysML 1.2 in June 2010 and SysML 1.3 in June 2012. At the time of writing this chapter, most tools run version 1.2 and for this reason all diagrams and figures herein illustrate version 1.2.

3.2. The origins of SysML

The software world reached an agreement at the end of the 1990s on a unified modeling language: UML. In 2003, the International Council on Systems Engineering (INCOSE) decided to make UML the common language for SE. At this time, UML already contained a number of indispensable diagrams, such as diagrams for sequences, states and use cases. Work on the new version UML 2, which began at OMG at around the same time, resulted in a definition for a modeling language that very nearly met the needs of systems engineers, with some important improvements in activity and sequence diagrams, as well as the development of composite structure diagrams.

There was still a significant psychological barrier to the adoption of UML by the SE community: its "software" terminology! The possibility of extending UML, due to the concept of "stereotype", allowed the vocabulary to be adapted for system engineers. By eliminating the words "object" and "class" in favor of the more neutral term "block", that is by removing the terms in UML that are the ones most linked to computer science, and by renaming this modeling language, the OMG wanted to promote SysML as a new language, different to UML, which was enriched by virtue of being directly linked to it.

The OMG announced the adoption of SysML in July 2006 and the first version (SysML v1.0) became available in September 2007. A new specification (SysML v1.1) was made public in December 2008, and the current revision (SysML v1.3) was published in June 2012.

3.3. General overview: the nine types of diagrams

UML 2.0 proposed 13 types of diagrams. Some of these have been reused exactly as is, some have been modified and others have not been kept. SysML is organized around nine types of diagrams, which the OMG has divided into three broad groups (see Figure 3.1[1]).

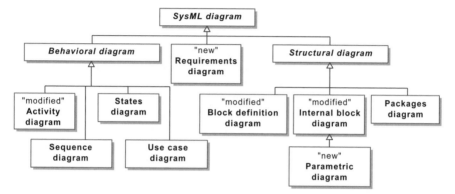

Figure 3.1. *The nine types of SysML diagrams [OMG 12]*

The first group is of four behavior diagrams:

– activity diagram (flow of actions and decisions);

– sequence diagram (vertical sequence of messages);

– state machine diagram (states and possible transitions of blocks, for example);

– use case diagram (system functionalities).

The second group consists of a single transverse diagram: the requirement diagram (showing the system requirements and the relationships between them).

The third group consists of four structural diagrams:

– Block definition diagram (showing the basic, static bricks, blocks, compositions, associations, attributes, operations, generalizations, etc.).

– Internal block diagram (showing the internal organization of a complex static element, in terms of its parts, ports, connectors, etc.).

– Parametric diagram (constraints and equations).

– Package diagram (logical arrangement of the model).

---

1 Most of the diagrams in this introductory chapter have been taken from the book by P. Roques [ROQ 09] or from OMG reference documents [OMG 12]. The case study diagrams in this book are referenced in the relevant sections.

The official presentation of how these different types of diagrams are organized is given in Figure 3.1. This figure also details the differences with UML by indicating the diagrams that have been modified or included unchanged.

In the rest of this chapter, we present the types of diagrams in their natural order of use: requirement, structure, dynamic and transverse.

3.4. Modeling the requirements

Taking the requirements into account in SysML involves several levels. SysML is innovative in comparison with UML in that it offers a requirement diagram that enables us to model the system requirements and, in particular, to then link these to the structural or dynamic elements in the model, as well as to requirements at the subsystem level. Nevertheless, we will begin this section by describing the use case diagram, which already exists in UML, and which is also very useful in describing the major functionalities required by the system and is the start of studying the functional requirements.

3.4.1. *Use case diagram*

The use case diagram (see Figure 3.2) is a schema that shows the use cases (ovals) linked by associations (lines) with their actors (stick-man icon, or equivalent pictorial representation). Each association simply indicates "participates in". This diagram type is strictly identical to its UML counterpart. A use case represents a collection of sequences of actions performed by the system, which produces an observable result that is of interest to a particular actor. An actor represents a role played by a human user or other system that interacts directly with the system in question (the subject), which is represented by the rectangle around the use cases.

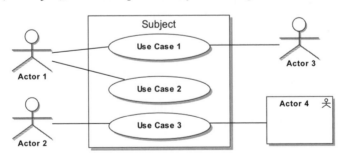

Figure 3.2. *A use case diagram*

Two or more actors may exhibit similarities in their relations with the use cases. We can express this by creating a generalized actor that models the aspects that are common to the different real actors. To refine the use case diagram further, SysML defines three types of standardized relations between use cases (see Figure 3.3):

– An inclusion relation, formalized by the keyword *include*: the basic use case necessarily and explicitly includes another use case, in an obligatory way.

– An extension relation, formalized by the keyword *extend*: the basic use case may implicitly include another use case. There will be a specific point within the primary use case at which the secondary use case may be invoked. This implicitly defined point is referred to as the *extension point*.

– A generalization/specialization relation (triangular arrow): the "children" use cases inherit the description of their common parent but may include additional specific interactions.

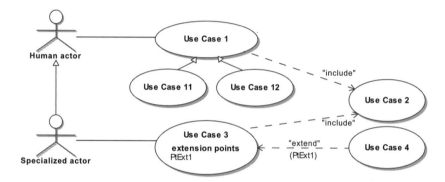

Figure 3.3. *Advanced concepts in the use case diagram*

For a concrete illustration of use case diagrams in the case study in this book, see Figures 4.3 and 7.1.

3.4.2. *Requirement diagram*

The requirement diagram offers a graphical representation of the requirements in the model. The two basic properties of a requirement are, first, a unique identifier (which gives traceability with the architecture, etc.), and second, a textual description (see Figure 3.4).

It is usual to define other properties of the requirements. The following list is not exhaustive:

– priority (e.g. high, medium or low);

– source (e.g. customer, marketing, technique and legislation);

– risk (e.g. high, medium or low);

– status (e.g. suggested, validated, implemented, tested and delivered);

– verification method (e.g. analysis, demonstration and test).

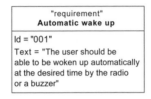

Figure 3.4. *Example of a requirement [ROQ 09]*

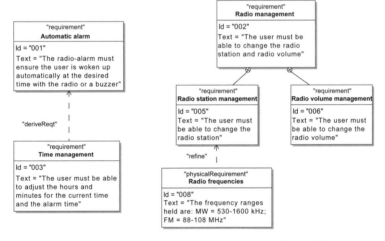

Figure 3.5. *Example of a requirement diagram [ROQ 09]*

The requirements can be connected to one another by relations of containment, refinement or derivation relations (see Figure 3.5):

– Containment (line terminated by a bounded cross on the side of the "parent" or "containing" use case) makes it possible to deconstruct a composite requirement into several single requirements, which are then easier to trace with regard to architecture and tests.

– Refinement (refine) consists of adding precision: for example, quantitative data.

– Derivation (deriveReqt) consists of connecting requirements from different levels: for example, connecting system requirements to requirements at the subsystem level. This generally involves making some choices in terms of architecture.

Through the different stages of a project, this requirement diagram can be used to make the connection between the requirements and all other types of SysML elements, by way of various kinds of relations[2] (see Figure 3.6):

– The relation between a requirement and a behavioral element (use case, state diagram, etc.) is denoted by the keyword *refine*.

– The relation between a requirement and an architecture block is denoted by the keyword *satisfy*.

– The relation between a requirement and a test case is denoted by the keyword *verify*.

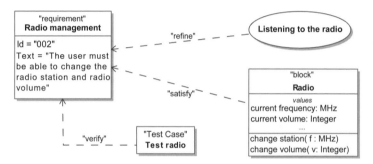

Figure 3.6. *Example of relations with the requirements [ROQ 09]*

For a concrete illustration of requirement diagrams in the case study in this book, see Figures 4.5, 4.14, 4.17 and 4.20.

3.5. Structural modeling

To model the static aspects of the system, SysML offers two main types of diagram: the block definition diagram (*bdd*) and the internal block diagram (*ibd*). The SysML block is a basic brick for modeling the structure of a system. We can use it to represent physical entities (complete system, subsystem or elementary component), and also for logical or conceptual entities. The block can also be used to describe flows (of data and/or control) that pass through a system. Blocks can be broken down and may exhibit behavior.

---

2 Another graphical representation offered by SysML consists of making the traceability relations obvious by using a note attached to a requirement, or by adding attributes or extra compartments. Another possibility is to use a table format.

3.5.1. *Block definition diagram*

The block definition diagram is used to represent blocks, their properties and their inter-relations. In a *bdd*, a block is represented graphically by a rectangle subdivided into compartments (see Figure 3.7). The name of the block appears at the top, and is the only obligatory compartment. All the other compartments have labels that indicate what they contain: values, parts, etc. Properties are the basic structural characteristics of blocks. The properties of the blocks may be of several types:

– Value properties describe quantifiable characteristics in terms of value types (range of values, dimensions and optional units).

– Part properties describe the decomposition hierarchy of the block in terms of other blocks.

– Reference properties describe relations of association or simple aggregation with other blocks.

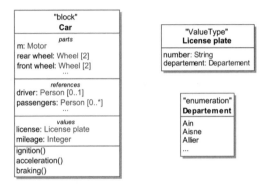

Figure 3.7. *Block notation in the* bdd

To define the types for value properties, SysML offers value types. In the example in Figure 3.7, the value type "License Plate" makes it possible to create a type that we can then reuse in the block "Car".

There are two main types of relations between blocks: association (with its two particular cases, aggregation and composition), and generalization (see Figure 3.8).

Association is a static, enduring relation between two blocks. An indication of multiplicity should appear at each of its two ends. It specifies, as an interval, the number of instances[3] that can participate in a relation with an instance of the other block in the context of this association. A unidirectional association has an arrow pointing toward the block that is being referred to.

---

3 An instance is an example of a certain block that has its own identity.

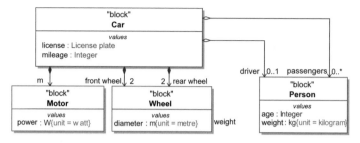

Figure 3.8. *Examples of relations in the* bdd

When the association is a "containment" type, we use the aggregation/composition relation. The interblock relation of a composition, whereby one block represents the coherent whole, and the others its constituent parts, can be represented graphically by a solid diamond on the side of the "container" block. For example, elimination of an instance of the container will result in the elimination of all the instances contained therein (cascade or "domino" effect). The aggregation relation (empty diamond) is far less strong than the composition relation (solid diamond) in the sense that the instances can exist independently of one another.

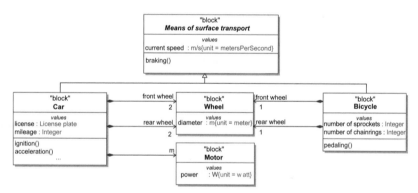

Figure 3.9. *Examples of generalization in the* bdd

Blocks can be organized in a classification hierarchy. This is often intended to factorize properties that are common to several blocks (values, parts, etc) in a generalized block. Specialized blocks "inherit" the properties of the generalized block and may have specific additional properties. Generalization is represented graphically by a triangular arrow that points to the generalized block (see Figure 3.9).

For a concrete illustration of block definition diagrams in the case study in this book, see Figures 4.1, 4.9, 4.10 and 4.11.

3.5.2. *Internal block diagram*

The internal block diagram (*ibd*) describes the internal structure of a block in terms of parts, ports and connectors. It is important to note that we can represent several levels of decomposition in a single *ibd*.

A composition relation in a *bdd* can be represented with an *ibd*. The *ibd* framework therefore represents the encompassing block. It provides the context for all the elements of the diagram. Each end of the composition relationship that exists in the *bdd* is presented as a block (known as a part) in the framework of the *ibd*. The name of the part is in the form: part_name: block_name [multiplicity] (see Figure 3.10). The multiplicity (1, by default) can also be represented in the upper right corner of the rectangle. Associations and aggregations that are outside the encompassing block are represented in a similar way to compositions, except that the line surrounding the block is dashed.

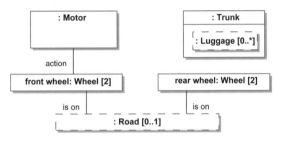

Figure 3.10. *Internal block diagram*

The connector is a structural concept used to connect two parts and to provide them with the opportunity to interact (the term "action" in Figure 3.10). The end of a connector can have a multiplicity, which describes the number of instances that can be connected by the lines described by the connector.

The port concept in the internal block diagram also enables descriptions of the logic behind the connection, services and flows between blocks. Ports define the points of interaction *provided* and *required* between blocks. A block can have several ports that specify different interaction points. There are two types of ports:

– Standard port: this type of port facilitates the description of logical services between blocks, through interfaces that bring operations together (this is the classic notion of application programming interface (API) in programming). These are represented by empty squares.

– Flow port: this type of port is new to SysML and can be used to produce a representation of the physical flows between blocks. The nature of such flows ranges from fluids, to energy, to data.

Flow ports are atomic (one flow) or non-atomic (aggregated flows of different types). An atomic flow port specifies only one type of flow that enters, or exits (or both), and the direction is simply indicated by an arrow inside the square, which represents the port. Its type can be given by a block or a value type, which represents the type of element that can flow into or out of the port (see Figure 3.11).

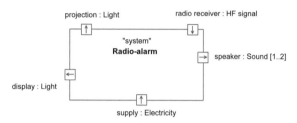

Figure 3.11. *Atomic flow ports [ROQ 09]*

When an interaction point has a complex interface with several flows, the corresponding flow port should be modeled as a composite (or non-atomic) flow port. In this case, the port type is given by a flow specification. This flow specification should be defined in a *bdd*. This includes several flow properties, each one of which has a name, a type and a direction. A composite flow port is shown graphically by two opposing chevrons (< >) drawn inside the port symbol. The conjugated flow port (prefixed by "~" or represented in black) takes the opposite inputs/outputs to those in the flow specification (see Figure 3.12).

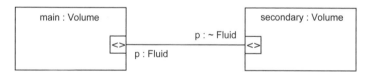

Figure 3.12. *Non-atomic and conjugated flow ports*

The interface on a standard port specifies the operations that the block provides. The realization relation that operates between a block and an interface is drawn as a dotted generalization. If the interface is represented by a simple circle, the realization becomes a simple line (see Figure 3.13). An interface that is required specifies the operations that the block needs to carry out in order to behave in the correct manner. The use relation between a block and an interface is drawn with a dashed arrow. In the condensed graphical notation, use becomes a simple line and the interface is represented by a semi-circle (see Figure 3.13).

For a concrete illustration of internal block diagrams in the case study in this book, see Figure 4.12.

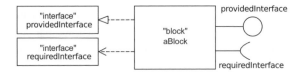

Figure 3.13. *Standard port with provided and required interfaces*

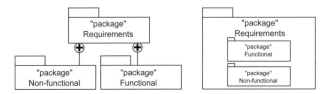

Figure 3.14. *Containment between packages*

3.5.3. *Package diagram*

The package diagram shows the logical organization of the model in a tree form as well as the actual dependency relations between packages. A package is a namespace for the elements that it contains. There are therefore two possible types of relation between packages: containment and dependency.

The containment relation can be represented in two ways:

– a bounded cross on the side of the container ($\oplus$);

– a graphical nesting of packages contained inside the encompassing package (see Figure 3.14).

Depending on how the model is organized and the structural choices, the elements of the different packages are often connected to one another. We have seen, for example, that blocks can be connected by associations, compositions, generalizations, etc. These relations between elements induce dependency relations between the overarching packages, represented by dashed arrows (see Figure 3.15).

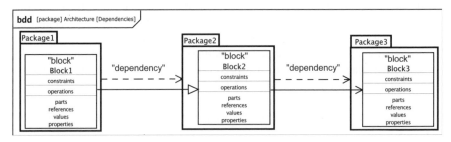

Figure 3.15. *Relations between blocks and dependencies between packages*

A view is a type of package used to show a particular perspective of a model, such as security, or performance (see Figure 3.16). A view corresponds to a viewpoint.

A viewpoint represents a particular perspective that specifies the contents of a view. A viewpoint contains properties that are standardized by SysML, as illustrated by the example in Figure 3.16:

– Concerns: the concerns of the stakeholders involved.

– Languages: the languages used to present the view.

– Methods: the methods used to establish the view.

– Purpose: the reason for presenting this view.

– Stakeholders: the parties who have an interest in the view.

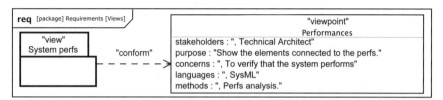

Figure 3.16. *Graphical representation of a viewpoint [ROQ 09]*

For a concrete illustration of the package diagrams for the case study in this book, see Figure 6.1.

3.6. Dynamic modeling

The diagrams in this section are concerned with behavioral aspects of the system. The use case diagram is classed under this heading in the OMG documentation, and is presented in section 3.4. Three complementary diagrams enable us to model the system's behavior: the sequence diagram models the exchanges between elements within an interaction, the state machine diagram models the behavior of an element in the form of states and transitions; and the activity diagram expresses the governing logic and inputs/outputs of a complex process.

3.6.1. *Sequence diagram*

The SysML sequence diagram is taken unchanged from UML 2. It shows the vertical sequence of messages that pass between elements (lifelines) in an interaction.

A lifeline has a name and a type. It is represented graphically by a dashed vertical line.

A message represents a unidirectional communication between lifelines, which triggers an activity on the part of its addressee. A synchronous message (transmitter blocked while awaiting response) is represented by a solid arrow while an asynchronous message is represented by a hollow arrow. An arrow that loops (reflexive message) is used to represent an internal behavior. The dashed arrow represents a return. This means that the message in question is the direct result of the previous message (see Figure 3.17).

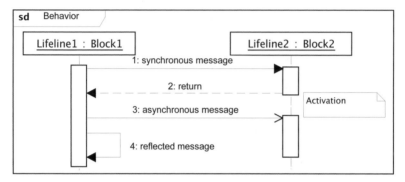

Figure 3.17. *Basic notation for the sequence diagram [ROQ 09]*

SysML offers a very useful notation: the combined fragment. Each fragment has an operator and can be divided into operands. The main operators are:

– loop: the fragment can be executed several times, and the guard condition makes explicit the iteration;

– opt: optional. The fragment is only executed when the condition provided is true;

– alt: alternative fragments. Only that fragment whose condition is "true" will be executed (see Figure 3.18).

A sequence diagram can also refer to another sequence diagram using a rectangle with the keyword *ref*. This notation is very practical for modularizing sequence diagrams, and for creating graphical hyperlinks, which can be used by modeling tools.

SysML also allows time constraints to be added to the sequence diagram. There are two types of limits:

– Duration constraints, which indicate the constraint on the exact, minimum or maximum interval between two events.

– Time constraints, which enable positioning of labels associated with particular times in the scenario at the level of certain messages and connect them to one another (see Figure 3.19).

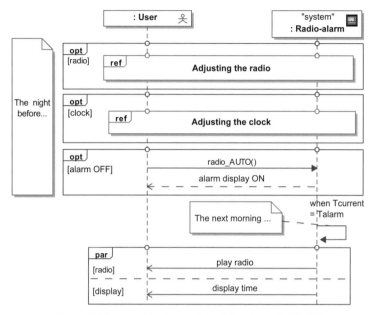

Figure 3.18. *Example of a combined fragment [ROQ 09]*

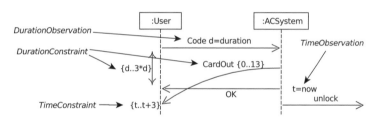

Figure 3.19. *Example of time constraints [OMG 12]*

For a concrete illustration of sequence diagrams for the case study in this book, see Figures 4.4 and 4.18.

3.6.2. *State machine diagram*

SysML reuses the well-known concept of finite state machines (already available in UML), which deals with the lifecycle of a generic instance of a particular block (in the sense described in the structural part above) in all possible cases for its interactions (with the other blocks). This local view of a block, which describes how it reacts to events as a function of its current state and how it changes to a new state, is graphically represented by a state machine diagram.

A state represents a situation in the life of a block during which:

– it satisfies a particular condition;

– it performs a particular activity;

– it waits for a particular event.

A block goes through a succession of states during its existence. A state has a finite duration, which varies according to the life of the block, and in particular is a function of the events that happen to it.

A transition describes the reaction of a block when an event happens (generally, the block changes state, but not always). As a general rule, a transition has a trigger event, a guard condition, an effect, and a target state. A transition can specify an optional behavior that the block performs when the transition is triggered. This behavior is called "effect": this can be a simple action or a sequence of actions. The do-activities have a certain duration, are interruptible and are associated with states.

In addition to the succession of "normal" states, which corresponds to the lifecycle of a block, the state machine diagram also includes two pseudo-states:

– The initial state of the state machine diagram corresponds to the creation of an instance.

– The final state of the state machine diagram corresponds to the destruction of the instance.

The basic notation for the state machine diagram is given in Figure 3.20.

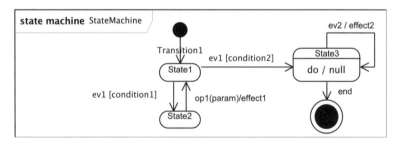

Figure 3.20. *Basic notation of the state machine diagram [ROQ 09]*

A composite state (also known as a superstate) encompasses several exclusive substates. Hence, we can factorize transitions that are triggered by the same event and lead to the same target state, by specifying the particular transitions between the substates. Another way of representing a composite state is to add a dumbbell-shaped symbol underneath on the right of a rectangle with rounded corners, then to describe the transitions between these substates in another diagram. In this way, we can deconstruct the state hierarchy, while keeping each level readable and relatively simple. We can also reuse state machines described elsewhere.

A composite state can also contain concurrent regions that need only be separated by dashed lines. Each region can then be named (optional). Each region contains its own states and transitions. The regions are known as concurrent because they can evolve in parallel and independently of one another.

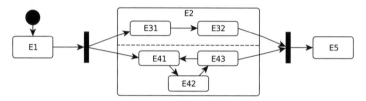

Figure 3.21. *Example of concurrent regions*

The language of the SysML state machine diagram contains many more advanced concepts, which will not be presented here since they are used much less often (entry, exit, internal transition, etc.).

For a concrete illustration of state machine diagrams for the case study in this book, see Figures 4.2, 4.13 and 4.15.

3.6.3. *Activity diagram*

The activity diagram represents the flows of data and control between actions. This diagram was already presented in UML 2 and has a few additions, the most important being concerned with the modeling of continuous flows.

Mainly, it is used to express the control and entry/exit logic. The basic elements of the activity diagram are as follows:

– actions;

– control flows between actions;

– decisions (also called conditional branches);

– a beginning and one or more possible endings.

The action is the fundamental behavioral specification in SysML. It represents a treatment or a transformation. Actions are contained in activities, which provide their context. Sequencing during the execution of activity nodes is controlled by the flow. The control flows are simple arrows that connect two nodes (actions, decisions, etc.).

The activity diagram also allows the use of object flows (connecting an action and an object consumed or produced). A decision is a structured control node that represents a dynamic choice between several conditions that must be mutually exclusive. It is represented by a diamond that has an arrow entering, and several arrows exiting it (see Figure 3.22).

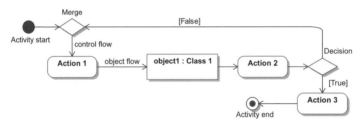

Figure 3.22. *The bases for the activity diagram*

A fork is a structured control node that represents a parallel branch. It is represented by a horizontal or vertical bar that has one arrow entering and several exiting it. The fork duplicates the token going in on each flow going out. The tokens on the exiting arrows are independent and concurrent. A join is a structured control node that represents synchronization between actions (meeting). It is represented by a horizontal or vertical bar with an arrow entering it and several exiting it. The join does not produce its own exit token since a token is available on each flow coming in.

Another interesting control node is the flow final. In contrast to the end of activity, which is global to the activity, the flow final is local to the flow concerned and has no effect on the overarching activities.

There is one last control node: merge. This is the inverse of the decision: the same diamond symbol, but this time with several flows going in and only one going out.

The activity diagram serves to define not only the sequence of actions to be performed, but also the flows that are produced, consumed or transformed during the course of executing these actions.

In order to do this, SysML offers concepts for the object flow and the input and output pins. An action handles the tokens on the input pins. These tokens, which can be represented by blocks, are consumed or transformed by the action, then placed on the output pins to be passed to other actions.

Activities can be reused through call actions (callBehaviorAction). The call behavior action is represented graphically by a fork to the right of the action box (see Figure 3.23), as well as by the string: action name: activity name. SysML offers plenty of other concepts and notations, such as the interruptible region, the expansion region or *stream-type* flows.

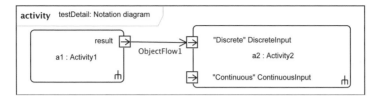

Figure 3.23. *Actions with pins, object flow and object node*

To enable modeling of continuous systems, SysML builds on UML 2 with the added possibility of characterizing the nature of the rate at which the flow circulates: continuous (e.g. electric current fluid) or discrete (e.g. events requests). We use stereotypes for this: continuous and discrete (see Figure 3.23). By default, a discrete flow is assumed.

For a concrete illustration of activity diagrams for the case study in this book, see Figures 4.7 and 4.8.

3.7. Transverse modeling

Independently of the structural or behavioral character, SysML adds a certain number of general concepts to UML 2.

SysML makes it possible to use the graphical notes from UML on all types of diagram (in the form of a Post-it note). Two particular keywords have been added to represent (see Figure 3.24):

– the problem to be solved (problem);

– the justification (rationale).

3.7.1. *Parametric diagram*

The parametric diagram enables constraints on system parameter values to be represented, such as performance, reliability and mass. This new diagram thereby provides a valuable support for systems analysis studies.

Each constraint is first defined by parameters as well as a rule, which describes the evolution of parameters with regard to one another. A constraint is represented by a block with a stereotype denoted as "constraint". We must therefore declare constraints in a *bdd*, as for more standard blocks (see Figure 3.25).

These constraints are then applied to the parametric diagram, which is a specialized internal block diagram, in order to gather them together and to draw the

connections between them. The constraint properties are represented differently to
the part properties in the *ibd*: these are rounded rectangles. The formal parameters
for these equations are represented by ports and can hence be connected to one
another (see Figure 3.26).

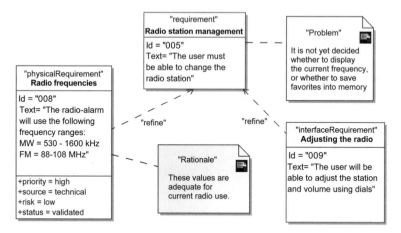

Figure 3.24. *Predefined notes for problems and rationales in SysML [ROQ 09]*

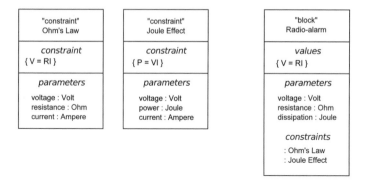

Figure 3.25. *Example of constraints declaration [ROQ 09]*

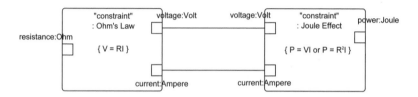

Figure 3.26. *Example of a constraints network [ROQ 09]*

The constraint is then instantiated by connecting values that belong to the model blocks to the formal parameters: this is the notion of value binding.

3.7.2. *Allocation and traceability*

Allocation is a mechanism widely used in SE to interconnect different types of elements – for example to connect an object flow in an activity diagram to a connector in an internal block diagram (*ibd*), or to connect an action to a block. These relations are often present in a classical model, but only in the form of comments or notes, and hence not very visible. The significant contribution of SysML to the allocation mechanism is in rendering visible, and useable, all these interconnections.

Since the allocation mechanism is very widely used, SysML has not added a specific diagram, but rather it has added a mechanism with several graphical representations available in numerous diagrams.

The first graphical representation is given by a dashed arrow with the keyword allocate (see Figure 3.27).

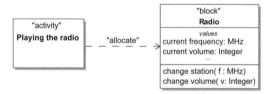

Figure 3.27. *General notation for allocation [ROQ 09]*

A second way of doing this consists of showing the relations in an extra compartment. The element indicated by the relation allocate has a property allocatedFrom, while the target element has a property allocatedTo (see Figure 3.28).

Figure 3.28. *Notation for allocation using additional compartments [ROQ 09]*

A third way of representing these properties is to include them in an attached note (see Figure 3.29).

In the same way as for requirements, SysML advocates being able to represent allocations in tabular form. For a concrete illustration of allocations drawn from the case study in this book, see Figures 4.16 and 4.19.

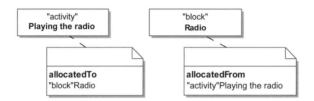

Figure 3.29. *Notation for allocation by adding notes [ROQ 09]*

3.8. Environment and tools

Among the main tools currently in use for SysML modeling, we cite Atego/ARTiSAN Studio, IBM/Rhapsody, No Magic/MagicDraw, ModelioSoft/ Modelio, Sparx Systems/Enterprise Architect, Papyrus (open source), Topcased (open source).

Note that this list of tools changes regularly. Their detailed functionalities (publishing, simulation, documentation generation, gateways into other languages and tools, etc.) are also frequently updated[4].

3.9. Conclusion

SysML is defined as a modeling language for the analysis and specification of complex systems. It is based on UML and recycles some of the important diagrams from UML. While UML is well adapted for SE, it does have shortcomings when applied to SE, such as the modeling of requirements or continuous flows.

Through reducing the number of diagrams offered by UML, while still extending its modeling capability, SysML aims to address the complete range of issues in SE.

3.10. Bibliography

[FRI 11] FRIEDENTHAL S., MOORE A., STEINER R., *A Practical Guide to SysML, 2nd ed.*, OMG Press, 2011.

[HOL 07] HOLT J., PERRY S., *SysML for Systems Engineering: The Emperor's New Modelling Language*, IET, 2007.

---

4 For an illustration of the compared functionalities of some of these tools, see en.wikipedia.org/wiki/List_of_Unified_Modeling_Language_tools

[OMG 05] OMG, Unified Modeling Language Specification 2.0: Superstructure, 2005. Available at OMG doc. formal/05-07-04.

[OMG 12] OMG, Official site of OMG, 2012. Available at www.omgsysml.org/.

[ROQ 09] ROQUES P., *SysML par l'exemple*, Eyrolles, 2009.

[WEI 08] WEILKIENS T., *Systems Engineering with SysML/UML: Modeling, Analysis, Design*, OMG Press, 2008.

Chapter 4

Modeling of the Case Study Using SysML

4.1. Introduction

In this chapter, we discuss the SysML model of our case study, considering the requirements defined in Chapter 2 as a starting point.

This work will be carried out in four stages, in the following chronological order:

1) first, we will perform a system analysis (detailed in section 4.2): here, we will describe the context, revisiting, in the model, the textual requirements described previously; we will also identify the use cases of the system, which represent its main capabilities; and we will establish the traceability relationship between the requirements and the use cases;

2) second, we will produce the system design (detailed in section 4.3): first, we will detail the identified use cases with activity diagrams. Then, we will identify the business data that are either manipulated by the system or visible for its environment. Then, we need to build the logical architectural model with the use of block diagrams (definition diagrams and internal diagrams) and state diagrams for describing the system as a set of elements communicating with each other and providing services so that all these services put together implement the expected system capabilities. Finally, we will complete this phase by describing the physical architectural model in order to answer the question: which are the physical components that are required for building the product?

3) third, we will specify traceability and allocation links (detailed in section 4.4): the aim of this activity is to consolidate the different models elaborated throughout

Chapter written by Loïc FEJOZ, Philippe LEBLANC and Agusti CANALS.

the project. This consolidation is mainly done by creating traceability links along three axes – satisfaction links between the logical architectural elements and the earlier requirements, links allocating the elements of the functional model (the use cases) to the logical elements and links allocating the logical elements to the physical architectural elements;

4) finally, we will build the test model (detailed in section 4.5): this is the final activity in the modeling process. This phase aims to build a repository of testing cases via sequence diagrams and occasionally activity diagrams. These test cases will be used as acceptance texts when the system has to be delivered.

In practice, these modeling activities will be carried out in a single project. This project will contain all of the products of these activities, generally stored in a set of independent packages. The structure of this SysML model reflects the chronology of these activities:

– "Context" package: system context;

– "System specification" package: system specification (outside of its context) comprising:

  - "General needs" sub-package: use case and traceability with the requirements;

  - "Technical needs" sub-package: requirements that are either directly created into SysML or imported from textual specification documents, and satisfaction links between the design elements.

– "System design" package: logical and physical design of the system:

  - "Functional architecture" sub-package: detailed description of the use cases;

  - "Domain-specific data" sub-package: description of the domain data or business data;

  - "Logical architecture" sub-package: system design in logical blocks and business data;

  - "Physical architecture" sub-package: system design in physical blocks;

  - "Allocations" sub-package: set of satisfaction links between requirements and logical elements, allocation links between functional and logical elements and allocation links between logical and physical elements;

– "System test" package: testing cases for the system.

Let us remember that the chronology of activities proposed here is not part of the SysML, which is neutral in relation to processes. However, it reflects a set of practices that are well accepted in the industry. For more information on this topic, we recommend the reader to refers to [HAS 12, KAP 07, FIO 12].

This process must be adapted to the complexity and the criticality of the system, as well as to the organization including roles and competences of the various actors concerned, as well as the geographical distribution of the teams (local teams versus distributed teams). SE2 [SE2 11] provides numerous ideas as to how a model from implementation on a large scale can be organized. Our proposition is highly dependent on the nature of our case study: a system of average complexity having little interaction with the external world. However, our system being critical, we will complete this SysML modeling with other types of modeling that will be detailed in Chapters 5 and 6.

These activities are generally performed sequentially. However, there are frequent iterations on the different models in order to ensure their coherence. This coherence is partially guaranteed by the modeling tools, which are able to propagate local modifications throughout the entire model, such as the renaming of an element (a drawing tool, for example, could not automatically propagate this kind of modification).

Because of a lack of space, the modeling work presented in this chapter is not exhaustive; we are not making a real pacemaker here. We will therefore carry out our modeling efforts on the pulse generator (PG) only in its ambulatory mode, that is the mode where the pacemaker is useful to the patient, after the implantation and before the explantation here means, the pacemaker is pulled out from patient's body. We will not detail the other parts of the system such as the control monitor.

4.2. System specification

In this stage, we will describe the system context, identify when it is likely to be used and integrate the textual requirements in the model in order to establish the traceability between the requirements and use cases.

4.2.1. *Context*

The context of the system must be described through its structure and the main phases of its lifecycle, which is done via the "context" and "lifecycle" diagrams.

4.2.1.1. *"Context" diagram*

The objective of the context is to present the system in its environment. The people that interact with the system, such as the users and the operators are represented by actors. The other elements of the external world that also interact with the system are represented by blocks. The system is also represented by a block.

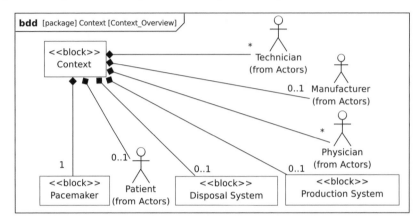

Figure 4.1. *Context of the system in SysML*

Details: in order to represent the entire domain of our case study, we have created a root block called "Context". Using compositions, this block now has all the necessary elements: the pacemaker (the system that is being studied) as well as the external people and systems that interact with it. These multiplicities are specified for the contained elements. The context is made up of the "Pacemaker" system for more than one patient (the patient does not participate in the phase of building the pacemaker), a set of physicians and technicians (who are not constantly present throughout the life of a pacemaker, but they are indeed present when the pacemaker, for instance, is being manufactured or when it is withdrawn), a manufacturer (that disappears from the context once the pacemaker starts being exploited) and finally a manufacturing system and a withdrawal system (which also appear and disappear depending on the phase the pacemaker is in). There are two different actors "Physician" and "Technician" which have been created for representing the two roles played by the medical staff in relation to the pacemaker: "Physicians" make surgical interventions, whereas "Technicians" only supervises the entire procedure.

Discussion: we have decided to represent the external systems by blocks whereas people are represented as actors, thus blocks and people are visually different. There are other practices that suggest we represent external systems as actors, just like the people are represented, but having a specific stereotype such as an "external system".

4.2.1.2. *Lifecycle diagram*

The lifecycle of the system (Figure 4.2) describes the different phases the system goes through in its lifetime, from its manufacturing to its withdrawal. We have used a state diagram attached to the "context" block to represent the system lifecycle.

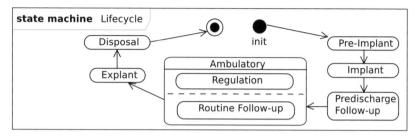

Figure 4.2. *System lifecycle*

Details: the "Regulation" and "Routine Follow-up" subphases are being executed in parallel in the "Ambulatory" phase. The provided specification does not give information on this, so we have decided that it was better for the patient to execute the two subphases in parallel, which is compatible with the technical constraints.

Discussion: the lifecycle described here enables us to explain in detail all of the possible phases of the pacemaker and thus, in order to come up with the best solution, to consider all of these phases (such as the withdrawal) without restricting ourselves to the ambulatory phase. The best solution for the ambulatory phase is not necessarily the best solution throughout the entire lifecycle. We meant to keep our analysis to a high degree of abstraction, so we have neither specified the conditions necessary for passing from one phase to another (from one state to another) nor the actions that would be executed.

4.2.2. *Requirements model and operational scenarios*

We will identify the main capabilities of the system by analyzing the textual requirements that were previously expressed, the capabilities being represented as use cases. As the name suggests, a use case corresponds to a set of similar scenarios that converge toward the same outcome and the same use. Some examples from everyday life are retrieving cash from an automatic teller machine (ATM) (whether that goes well or not), reserving a book in a library (whether it is available or not), etc. Typically, a use case will collect the nominal use scenarios and the alternative ones, whether they are failed or undesirable scenarios for the service that is expected of the system.

A difficulty frequently encountered in use case modeling is finding a good level of granularity. A use case is a complete service provided to the user, including all of the variations of that service. This is not an action of the system, such as emitting a beep in the case of an anomaly or printing a ticket at the end of a bank transaction. A use case must have a greater objective, which justifies the system being used by one of the actors, and this objective is achieved by answering questions such as "why does the

actor move?" and "what benefits (visible results) does the actor get from the system?".
We often estimate at maximum 20 the number of use cases that need to be processed
for a given modeling project.

Finally, it is customary to name the use cases using verbal phrases – such as "doing
something" – and using the vocabulary of the stakeholders (e.g. "registering the bank
transaction") and not the technical vocabulary that would rather reveal a design choice
(e.g. "registering the transaction in a database").

We have created a use case diagram (Figure 4.3) to identify the services provided
by the system. We have also identified the main services provided by the external
systems of the environment (production and retrieval). The use cases will then be
fully described in the system design phase.

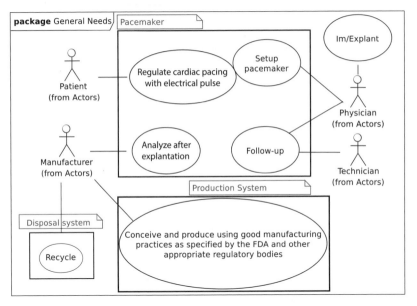

Figure 4.3. *Context use case*

We have also elaborated the sequence diagram "ConOps" (Figure 4.4) in order to
show the global operational scenario that corresponds to the nominal lifecycle of a
pacemaker.

4.2.2.1. *General needs diagram*

This SysML use case diagram (Figure 4.3) details the way in which actors use
systems – actors and systems having been defined in the context (Figure 4.1).

Details: the "Pacemaker" system supports four use cases provided to its different actors. In system design, we will detail the three use cases "Regulate cardiac pacing", "Setup pacemaker" and "Follow-up". The case "analyze after explementation" will not be detailed because it is not performed in the ambulatory phase of the system. These cases and the actors involved were identified from the textual requirements. Traceability will be described later.

We have identified the main capabilities of the production and withdrawal systems. They are not detailed in the rest of the book, but they would have to be detailed in an industry-related project.

Discussion: in a real modeling project, it is common to complement this diagram with descriptive forms, one form per case, respecting a standard format such as "Actors", "Preconditions", "Postconditions", "Nominal scenario" and "Exceptions". This documentation is designed for stakeholders who are not necessarily very familiar with the SysML. We will not do this here, because our objective is to make a model rather than to write textual documents.

4.2.2.2. Scenario "ConOps" nominal for the system

The sequence diagram in Figure 4.4 describes an operational scenario of the system's lifecycle in its environment.

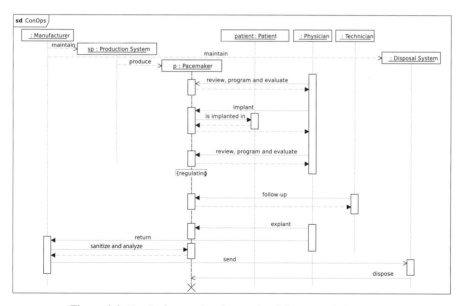

Figure 4.4. *Nominal operational scenario of the system in its context*

Details: the lifelines correspond to the "Pacemaker" system, to its actors and to the other interconnected systems. This sequence shows the nominal course of the life of a pacemaker, its creation, implantation, the ambulatory phase with its follow-up, the explantation, and at last its withdrawal. The "maintain" and "produce" interactions correspond to the creation of three systems: production, withdrawal and pacemaker. The other interactions between the lifelines correspond to synchronous messages with their execution areas and return arrows. The instance of the "Pacemaker" system disappears when the pacemaker is sent into the withdrawal system. The "regulating" condition added on the Pacemaker lifeline indicates the operational mode that the system is in after the operation "review, program and evaluate" has been called at the beginning of the sequence.

Discussion: the interactions have been represented by synchronous messages (the caller waits for the end of the called operation); we may also represent them through asynchronous messages (non-blocked caller), without an execution area or return arrows. However, in this context, the use of synchronous messages better represents the expected system behavior, because they impose the sequentiality of the events. For example, it is not possible to implement the system before carrying out the review and the assessment.

4.2.3. Requirements model

There are two questions that we must ask ourselves throughout the development of a model: "Is this a correct model from a SysML viewpoint?" and "Is this the right model from a functional point of view?"

To answer the first question, we must have a good knowledge of the SysML norm, and have acquired hands-on, "how to" knowledge. The modeling tools are also there to help us to build SysML models that are correct from the language point of view.

The solution generally answered for the second question is to specify the traceability between elaborated model and on the requirements. So, thanks to the traceability links, we can prove that the elements consituting the model fulfill functional expectations.

We will have to show that the model meets the criteria: all of the requirements are connected to elements of the mode. For the critical systems, we will also have to show that the model is necessary, that is it only does what is expected of it: all of the structuring elements (use case, logical blocks and software blocks) are connected to certain requirements. To establish this traceability, we must first redistribute the textual requirements initially expressed in the specification phase within the model. These requirements will be generally imported in a package that is unique to the root of the model. However, given the high number of requirements that might need to be imported, this root package can be cut out in several sub-packages, each sub-package representing a chapter or a section of the chapter from the document where the requirements are specified.

4.2.3.1. *"Technical needs: the pacemaker functions properly" diagram*

This SysML requirements diagram (Figure 4.5) shows the redistributed requirements of the specification document with regard to the correct functioning of the pacemaker. These requirements were connected to certain elements of the functional model. A requirement is modeled by the name, an identifying attribute and a text. The requirements can be structured between them via the "refine" dependance link[1] in order to express the fact that a requirement can be broken down into a set of more specific requirements.

In a first phase, we will look into the requirements. Then, we will tackle the traceability along with the use cases.

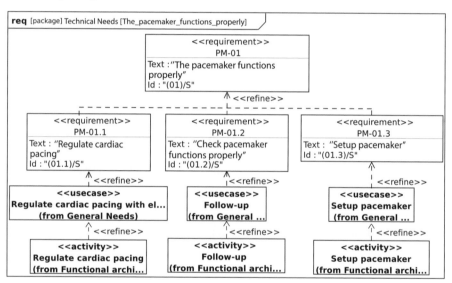

Figure 4.5. *Requirements for the correct functioning of the system*

Details: we have decomposed the PM-01 requirement, identified as (01)/S in the documentation, into three requirements PM-01.1, PM-01.2 and PM01-3 in the middle of the "refine" relation. To indicate that this decomposition into three sub-requirements is a partition (i.e. the three combined sub-requirements answer completely to the higher exigence), we have used the representation in the shape of a tree. If we had used direct arrows between a higher requirement and sub-requirements, this would have been to indicate that the decomposition is not a

---

1 The blog notes [FEJ 10, FEJ 11] justify the use of this type of links.

partition and that the PM-01 requirement is not entirely satisfied by the three visible sub-requirements.

Discussion: in order to hierarchically organize the requirements, we have used the "refine" dependance link. We could have used the composition instead. This choice reveals certain practices defined in the enterprise, the two uses being correct in our case study. One of the advantages of the "refine" is the possibility to express alternatives, as indicated in [FEJ 11].

The concepts of requirement and refinement in SysML does not allow us to verify neither the requirements nor their partitioning. In order to do this, we must apply differnt verification techniques which are discussed in the rest of this book.

4.2.3.2. *Needs – Use case traceability*

Once the requirements have been redistributed in the model, we must link them to the use cases identified in order to justify the fact that the analysis carried out is necessary and satisfactory. We can see this in Figure 4.5, which shows via "refine" relations the way in which the requirements are considered by the use cases.

Details: Figure 4.5 also shows, via "refine" links, which activity diagram details which use case, but this is rather the subject of section 4.3.

Discussion: we have made the choice to use the "refine" dependance between the use cases and the requirements. SysML also proposes the "satisfy" dependance, but we have preferred to save it to express the traceability between requirements and software architecture.

In the exact case of PM-01, the three technical requirements PM-01.X are each considered in a single use case, but we could have had M-N relations: a requirement can be connected to several use cases and conversely, a use case can be connected to several requirements.

To demonstrate that a model is necessary and satisfactory, we need every requirement to be linked to at least one use case and, conversely, that every use case is connected to at least one requirement. Since it is a costly effort, we will observe this for projects of critical systems.

4.3. System design

In the design phase of the system, we will detail the expected behavior for the use cases identified in the specification phase, and then propose an architecture of the pacemaker. The use cases are described in the functional part of the model, in the following paragraph.

The solution is divided into three parts:

– business data: this part of the model defines the data exchanged between the actors and external systems – which corresponds to the interfaces of the pacemakers – and the data consumed or produced by the pacemaker;

– software architecture: this describes the software components proposed for forming the pacemaker; the software offers functionalities that when put together, provide the services that are expected from that level of the system; at that stage, we do not seek to indicate if this or that component or this or that functionality is rendered by the software, the electronics, the mechanics, etc.

– physical architecture: this describes the physical design that is proposed as a solution and the hardware and software components are identified here.

The design phase ends with consolidation activities:

– the traceability between requirements and logical architecture: the requirements are linked to the logical components in order to demonstrate that the solution is well adapted to the needs;

– the allocation of the logical architecture to the physical one: the logical components are allocated to the physical components in order to show how the system can be deployed on the hardware environment.

These traceability and allocation activities will be detailed in section 4.4.

4.3.1. *Functional model*

In the specification phase, we have identified four use cases for our system. Only three of them concern us because they regard the ambulatory part of the pacemaker: "Setup pacemaker", "Regulate cardiac pacing" and "Follow-up". Each of these three use cases is illustrated via an activity diagram to specify the concatenation expected of the actions, which produces the concerned use case. Each of these diagrams was associated with its own use case by a "refine" link, which can be seen in Figure 4.5.

4.3.1.1. *"Setup pacemaker" activity diagram*

This SysML activity diagram (Figure 4.6) details the "Setup pacemaker" use case.

Details: as described in Chapter 2, the pacemaker is adjusted in two steps, once during the implementation and then a second time afterwards. The adjustments during the implementation consist of verifying the state of the battery, and then of programming the system and adjusting its parameters, and finally of taking the measurements. The adjustments after the implementation consist of retaking the measurements, consequently reprogramming the parameters and then drawing up reports and graphics on the basis of these data.

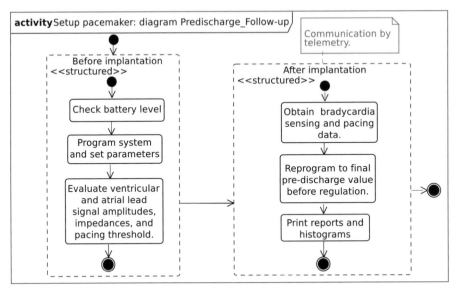

Figure 4.6. *"Setup pacemaker" activity diagram*

Discussion: the two large stages are modeled by two action blocks and each operation that must be carried out by medical personnel is represented by one action. We use control flows between the actions in order to specify their concatenation. We mention here "true" guards to signify that once the action is terminated, the next one is triggered automatically.

4.3.1.2. *"Regulate cardiac pacing" activity diagram*

This activity diagram (Figure 4.7) details the use case "Regulate cardiac pacing with electrical pulse".

Details: the functioning of this use case is limited to a loop in which we, sequentially, take measurements, control the impulse, adjust the impulse and finally send the impulse, and then the whole process starts over from the beginning. The control is done in relation with the mode and a value that is registered in a data store. The adjustment action will read this value in the data store, and then provide a value adjusted to the action of sending the impulse. In SysML, the data are moved around by object flows, connected to actions via oriented ports, whereas control flows only express a squence of actions. We have added two commentaries in order to specify the importance or the impact of certain actions.

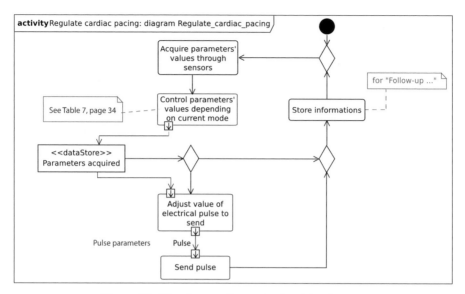

Figure 4.7. *"Regulate cardiac pacing" activity diagram*

4.3.1.3. *"Follow-up" activity diagram*

This SysML activity diagram (Figure 4.8) details the "Follow-up" use case.

Details: This service takes place in two stages: the first stage allows us to acquire the measurements via a wireless connection, while the second stage validates the changes if there have been any. During the acquiring stage, two action flows take place in parallel: the left flow triggers the connection, state verification, acquisition, analysis and finally the production of assessment reports; the right flow involves the updating of the parameters in order to regulate the pacemaker. The parallelism of these flows is represented by the use of a Fork-Join arrangement (vertical bars). After the acquisition stage, any potential changes in the functioning parameters must be validated. The histograms are erased at the end of the processing. We have also added a few commentaries in order to clarify certain actions.

Discussion: when we describe the behavior of a use case, the question of granularity and precision arises within the description of the actions. The level of detail must be coherent with the global objective that was fixed for the model. If we then wish to add details to an action, we will be able to use the commentaries.

4.3.2. *Domain-specific data*

This part of the model specifies the interfaces of the pacemaker with its environment, as well as the significant data that it manipulates (produced/consumed).

They come from a knowledge of the domain. They are described in the block definition diagrams via the following SysML concepts: interface, event, simple type, complex type represented by a block, datatype and value type. We will only present here two diagrams of this type, however the model contains others as well, such as, for example, the list of the modes supported by the pacemaker with their own types of associated values (and their values: min, max, increment, nominal value, tolerance) and the potential measurements.

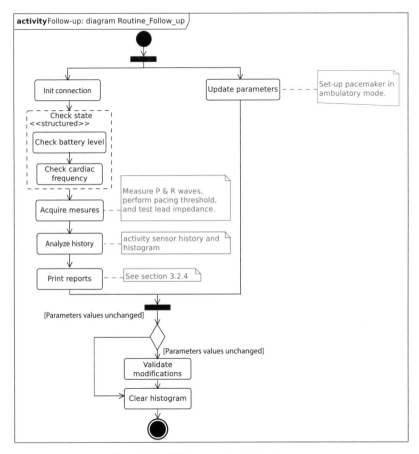

Figure 4.8. *"Follow-up" activity diagram*

4.3.2.1. *Domain-specific data diagram (PPM) frequency measurement*

Figure 4.9 is a block definition diagram that groups together a set of declarations: "unit", "dimension", "valauetype" and "datatype". They serve to specify the types of data that are connected to the frequency measurements in pulse per minute (PPM).

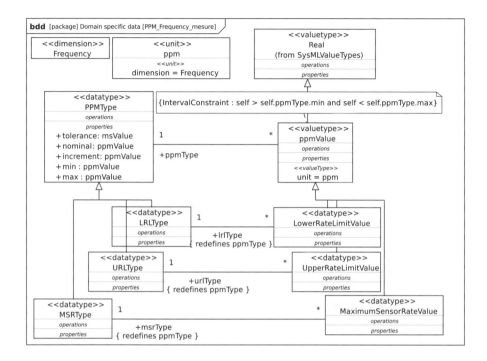

Figure 4.9. *Domain specific data diagram "PPM frequency measurement"*

Details: we have declared "Frequency" as a "dimension", and then the unity "ppm" as a "Frequency" dimension "unit". We will then declare "valuetypes" to represent simple values, and then "datatypes" to represent sets of simple values such as "LRLType", "URLType", and MSRType. These types will be used to characterize the data manipulated by blocks from the architecture of the system. We have also added a SysML constraint to specify the interval of the authorized values (between minimum and maximum). The "IntervalConstraint" expressed as an Object Constraint Language (OCL) invariant has been added to the "valuetype" and "ppmValue" in order to specify its domain of validity depending on the "min" and "max" fields of the associated "PPMType".

Discussion: the use of the "dimension", "unit", "valuetype" and "datatype" is the most precise SysML means for declaring the data. For a more abstract model, we would have preferred the use of basic types such as the integer, the string of characters or the chain of types that had not been formally described.

4.3.2.2. *Domain-specific data diagram "Time measurement"*

This SysML diagram for the definition of blocks (Figure 4.10) serves to specify the types of data connected to the time measurements for our pacemaker.

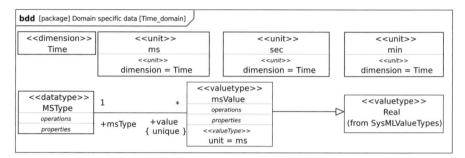

Figure 4.10. *"Time measurement" domain-specific data diagram*

Details: we have declared "Time" as a dimension, then the units "ms", "sec" and "min" as "units" of the "Time" dimension. We will then declare a "valuetype" to represent the simple value "msValue". This type will allow us to build a complex data such as the "MSType".

4.3.3. *Logical architectural model*

Logical architectural model objective is to propose an architectural solution. The system is decomposed into logical components that offer each a set of services. These components communicate between each other to trigger their individual services, which allows them to obtain at the level of the system the global services expected by its environment. This modeling is composed of block diagrams (definition and structure) to describe the architecture and the state diagrams to describe the behavior of each of the logical elements.

4.3.3.1. *"Logical Architecture" diagram*

This block definition diagram (Figure 4.11) describes the structure of the system in logical components.

Details: the system is decomposed into three first-level components: PG, device controller-monitor (DCM) and magnet. The "PG" block must be in itself broken down into three blocks: PGController, Battery and Lead, these blocks being terminal. The two constraints added in the "Lead" block allow us to specify the tolerated impediments, that is the constraints that come straight from the model of these requirements.

The blocks have ports in order to communicate data or events. Let us take the example of the "PG" block, which has:

– two "ElectricalFlow" ports, "HeartElectricalSignal" to represent the electrical current produced by the battery and the electrical impulse produced by the "Lead" toward the heart;

– a "PG Port" control port for receiving the events declared in the "IF DCM toward GP" interface and for sending the events declared in the "IF GP toward DCM" interface, this port will enable us to connect the PG to the DCM.

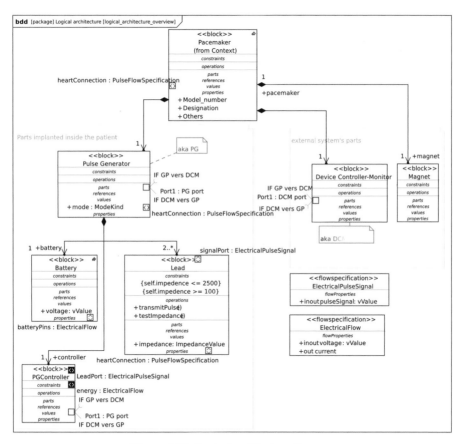

Figure 4.11. *Logical architecture diagram*

Two flow specifications "ElectricalPulseSignal" and "ElectricalFlow" are shown in the diagram in order to specify the content of flows transmitted by the ports of the "PGController", "Battery" and "Lead" blocks. The interfaces declared on the control ports are described in another diagram so the main diagram does not get cluttered.

This block definition diagram must be completed by a block structure diagram to establish the effective connection between the standard ports of the PG and DCM blocks. We have not included this diagram in this chapter, but it is present in the model.

In order to build a modular architecture while still respecting the levels of hierarchical structuring, the ports of the lowest components must also be found in all the levels back up on the hierarchy, until the highest level, that is the pacemaker system. One example is the "heartConnection" port of the "Lead" components, which we find all the way up to the "Pacemaker" block: it represents the existing connection between the system and the heart of the patient. The ports of the children blocks are connected to the ports of the parent blocks via internal block diagrams in order to ensure the communication between the different levels of decomposition. Figure 4.12 shows us one of these internal block diagrams. This describes the connections established between the control port of the PG block and the ports of its components PGController, Battery and Lead.

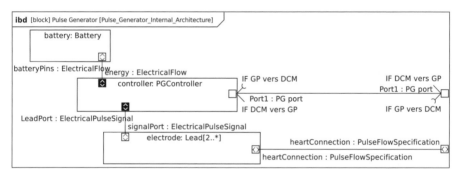

Figure 4.12. *Internal diagram of the Pulse Generator block*

Discussion: in order to facilitate the understanding of the logical architecture, in Figure 4.11, we put on the left-hand side the entire part of the system that has been implemented in the patient, and to the right-hand side the components that are outside the body of the patient. The structuring into components is classically represented by SysML compositions with the name of the roles and their associated multiplicity.

The blocks contain a certain number of properties such as *values*, *operations*, *constraints* or *parts*, which can sometimes become very important. It is up to the designers to choose the properties that they show in the block diagram so as not to clutter it pointlessly. We will, for instance, use several diagrams in order to show all of the properties of the blocks or make one diagram per block in order to showcase all of its properties.

4.3.3.2. *State diagram of the PG block*

This SysML state diagram (Figure 4.13) describes the evolutions and the reactions of the PG block throughout its use depending on the commands and events that it receives.

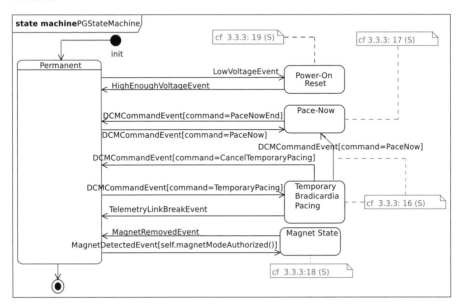

Figure 4.13. *State diagram of the Pulse Generator block*

Details: the PG has five modes that are directly identified starting from the specification of the requirements, each mode being represented by a state in the diagram. The transitions starting from or ending to the states "Pace-Now" and "Temporary Bradicardia Pacing" are controlled by DCM commands. The transitions toward or starting from the "Power-On Reset" mode are controlled by tension measurements. The transitions from or to the "Magnet State" are controlled by the detection of the magnet. In general, all of the mode changes go through the "Permanent" state, except the case where we can move directly from the "Temporary Bradicardia Pacing" state to the "Pace-Now" state, upon reception of the command "PaceNow".

Discussion: we have added references to the different paragraphs of the specification. A more rigorous manner would be to establish the traceability relations between the states of the PG automaton and the various requirements.

4.3.4. *Physical architectural model*

The architectural modeling of a complex system must be completed by a diagram showing the physical deployment of the system. This description is done via a SysML block definition diagram (BDD) that contains blocks representing the physical elements. We also show here the distribution between hardware and software, as well as the means of communication used, such as buses or protocols.

In the current case study, such a diagram would contain blocks for representing the electronic board that runs the software controlling the production of impulses, the electrodes, the battery, the wires connecting the battery to the electrodes, the wireless connection to the monitor, etc. We have not detailed it here because it does not bring any new SysML concepts in comparison with the logical architecture.

4.4. Traceability and allocations

The allocation and traceability activities together help us consolidate the different models carried out throughout the system engineering process. The traceability between the requirements and the architecture will allow us to demonstrate that the solution answers the respective needs and that our design choices are justified (indeed necessary). The traceability will be established between the requirements and the logical architecture. We will illustrate the traceability activity via two diagrams that can be seen in Figures 4.14 and 4.15. The allocation will enable us to establish a correspondence between the different architectures; how the functional is distributed on the logical and how the logical is used on the physical. We will illustrate the allocation activity via the diagram seen in Figure 4.16.

4.4.1. *"Technical needs: divers" traceability diagram*

This requirements diagram (Figure 4.14) shows the traceability relations between the requirements and system architecture elements, in this case, blocks and block attributes.

Details: the requirement of the first PM-03 level is first of all refined into three requirements, PM-03.X. The high-level requirement is satisfied by the "Pacemaker" block. The "id" attribute of the "Pacemaker" block satisfies the three requirements of the second level PM-03.X. The PM-03.1 requirement relies on the attributes "Model number" and "Designation" of the "Pacemaker" block. The PM-04 requirement is not refined and is satisfied by the "Lead" block.

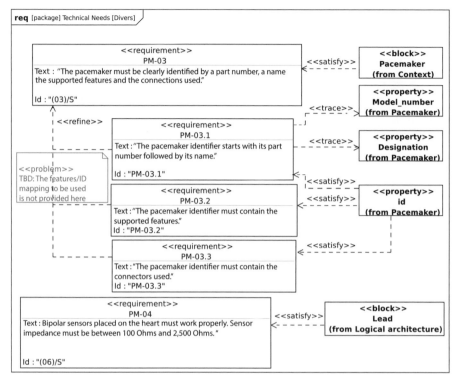

Figure 4.14. *Traceability diagram between requirements and blocks and logical data*

Discussion: a "problem" note was added to the "refine" decomposition between requirements to indicate that something is missing in the specification. It could have also been associated to the traceability relation between PM-0.3 and "Pacemaker".

The "satisfy" link indicates that the source block of the link should satisfy the target requirement. It is one of the objectives of the verification stage to check that this is done correctly.

4.4.2. Traceability diagram "technical needs: behavior of the pacemaker"

Figure 4.15 is a requirement diagram showing the traceability relations between the requirements and the behavior defined for the PG block.

Details: the different operational modes described in the specification are implemented by the states of the state machine of the "PG" block (Figure 4.13).

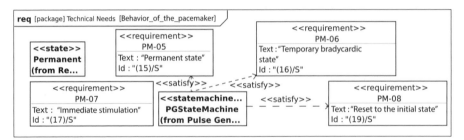

Figure 4.15. *Traceability diagram between requirements and behavior*

Discussion: the PM-07 requirement may be understood as the possibility to go from whichever PG state to the "Pace Now" state (the existence of a transition). This requirement is verifiable on the model, being expressed by an OCL constraint specifying that there is a transition that starts from every state toward "Pace Now". For instance, the tutorial [GAB 09] explains this in the Topcased tool.

4.4.3. *Allocation diagram*

Figure 4.16 is a block diagram that shows the allocation relations between the activities (functions) identified in a functional analysis and the blocks derived from the logical architecture.

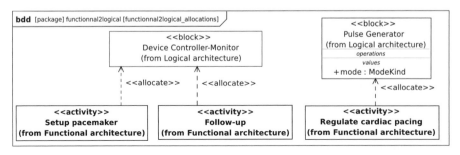

Figure 4.16. *Allocation between functional elements and logical elements*

Details: the "Regulate cardiac pacing" function (SysML activity) is allocated to the "PG" block, whereas the "Setup pacemaker" and "Follow-up" functions are allocated to the "DCM" block.

Discussion: these diagrams enable us to allocate the functions (represented by activities within the functional SysML model) to blocks of the logical architecture. An allocation between a function and a block creates, one or several operations in the target block in order to implement the function. In the end, all of the functions will have to be allocated to blocks, and this completeness control can be carried out by the SysML modeling tool.

4.5. Test model

The test model helps us define the acceptance criteria for the system. Generally, it is made up of sequence diagrams that describe operational scenarios in detail at the black-box system level: the interactions received via the lifeline of the system correspond to inputs or commands that must be applied to the system tested, whereas the interactions produced by the system correspond to observations that must be made by the test operator. Activity diagrams can also be used, as a complement or as an alternative, to describe the actions a test operator must make in order to test and validate the system.

One good practice is to also explain why a test needs to be done to validate the system. This is done by specifying the traceability links between test cases and requirements.

4.5.1. *Traceability diagram "system test: requirements verification"*

The requirements diagram seen in Figure 4.17 shows which requirements are tested by the test games. A requirement can be verified by several test cases and conversely a test game can verify several requirements.

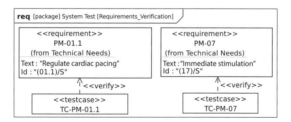

Figure 4.17. *Requirements – test games traceability*

Details: the TC-PM-01.1 test case tests the PM-01.1 requirement and the test game TC-PM-07 tests the PM-07 requirement.

4.5.2. *Sequence diagram for the test game TC-PM-07*

The sequence diagram in Figure 4.18 shows the sequence of inputs and the outputs that must be observed to have a successful TC-PM-07 test.

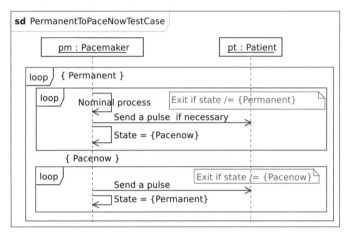

Figure 4.18. *Sequence diagram for the test case TC-PM-07*

Details: the test is being carried out between the "Pacemaker" system and the patient (who is an actor). It consists of a general loop allowing us to execute the test as many times as we want. This loop contains, in itself, two other loops, one is in the "Permanent" mode and the other in the "Pace now" mode.

Discussion: these diagrams allow us to specify the tests that will be used to validate the delivered system. This is interesting from two points of view. First, they will force us to find how to deal with the tough points, the technical points and the functional points – performances, robustness, etc. Second, they will allow us to create the validation plan, either in an automatic way (if the tool allows us to do so) or manually.

Of course, if we wish to do it exhaustively, the number of tests will probably be very (too) important and difficult to manage. We must therefore find a good compromise between the efforts invested and the expected test coverage, and we must also limit ourselves to relevant tests. This remains a difficult point in a SysML modeling project and can be solved with the help of tools such as graph analyzers or property checkers.

4.5.3. *Diagrams presenting a general view of the requirements*

Figures 4.19 and 4.20 are requirement diagrams showing the links between a requirement and elements of the model.

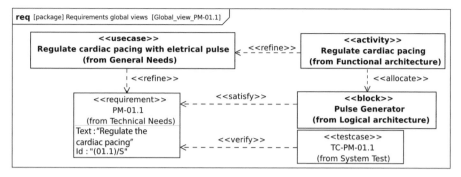

Figure 4.19. *Diagram showing the TC-PM-01.1 requirement*

Details: PM-01.1 is satisfied by the "PG" block. It is refined by the use case "Regulate...", itself refined by the "Regulate..." activity, which is allocated to the PG block. Finally, it is verified by the test scenario "TC-PM-01.1".

Discussion: these are optional diagrams. Indeed, if we wish to be exhaustive, we must make as many diagrams as there are requirements, which might mean hundreds. However, they are interesting because they enable us to represent a requirement in its ecosystem (relations established with the functional and logical parts of the model) on one diagram, thus facilitating navigation within the entire model. We will therefore only consider the major requirements. We must note that these diagrams do not add any new information, but represent the information defined in several diagrams in one place. This is the advantage of the modeling in comparison to a simple collection of drawings. Some tools are otherwise capable of recreating this type of diagram automatically by analyzing the complete model.

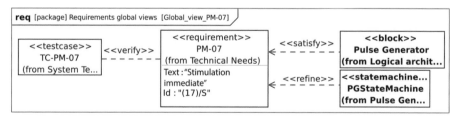

Figure 4.20. *Diagram showing the TC-PM-07 requirement*

4.6. Conclusion

The model-driven engineering of the pacemaker system has helped us to elaborate a global model that combines different design views as follows:

– requirements identified starting from the specification document;

– functional analysis for identifying the expected use cases;

– functional architecture for detailing the use cases;

– logical architecture for describing the logical solution based on communicating blocks;

– physical architecture for showing the use of the solution on physical components;

– test plan for the system acceptance.

Consolidation activities are carried out to ensure that all the model views are coherent with one another, that is:

– there is a traceability relationship between the different views: the level of requirements, the functional view and logical view;

– allocation links are created between the functional, logical and physical architectures;

– traceability links are created between acceptance tests and requirements.

This set of models exhaustively and rigorously describes the proposed solution for the needs that must be met. This modeling will be referenced in the next design stages. One of the main benefits is solutions with different levels of detail and formalization, from the simplest ones with the use cases to the most complex ones with ports, interfaces and automata. This facilitates communication and understanding for all stakeholders, from the client to the supplier, without requiring a large amount of SysML expertise.

Modeling also brings two other benefits: maintainability and scalability of the solution. Indeed, it is relatively easy to upgrade it along with future changes. For example, if some requirements are changed, thanks to their traceability links, we know quickly and precisely which functional elements, whether logical or physical, are being affected. It is also easy to upgrade the test plan by reconsidering the sequence diagrams in connection with the modifications done in the logical architecture: the interactions between the lifelines must rely on the operations and ports of the blocks.

SysML finally allows us to carry out risk analysis, performance analysis, cost analysis, design trade up analysis and safety analysis, by using the "View", "ViewPoint" or parameter diagrams. It is worth noting that the latter can be executed via dedicated tools such as mathematical problem solvers.

4.7. Bibliography

[FEJ 10] FEJOZ L., "How to link SysML requirements?", 2010.

[FEJ 11] FEJOZ L., "Solution space exploration with SysML requirements", 2011.

[FIO 12] FIORÈSE S., MEINADIER J., *Découvrir et comprendre l'ingénierie système*, AFIS, Editions Cépaduès, 2012.

[GAB 09] GABEL S., Topcased OCL Tooling Tutorial, 2009.

[HAS 12] HASKINS C., SE HANDBOOK WORKING GROUP, *INCOSE Systems Engineering Handbook: Version 3.2.2*, International Council on Systems Engineering, 2012.

[KAP 07] KAPURCH S., *NASA Systems Engineering Handbook*, DIANE Publishing Company, 2007.

[SE2 11] SE2 CHALLENGE TEAM, *Cookbook for MBSE with SysML*, 2011.

Chapter 5

Requirements Analysis

5.1. Introduction

In the early stages of the lifecycle of a complex system, gathering the requirements plays a key role in ensuring the phases of analysis, design, implementation and maintenance. The traceability of requirements throughout the system's lifecycle has thus become a major preoccupation for system engineers. The same engineers acknowledge the importance of a requirements analysis being carried out at the earliest stages of a system's lifecycle.

Besides the cross rereading that enables us to assess how coherent requirements are between one another, there is another essential type of analysis that enables us to detect design errors very early in the process and thus helps us to reduce the development as well as the testing costs of the system: checking the design model against the requirements. In a SysML context, this type of checking weighs an architecture of block instances and the behaviors of the entities that compose it, with the requirements being expressed in the requirements diagrams and in the use cases.

To check a design against its requirements, we need architecture diagrams and behavior diagrams that are both executable and non-ambiguous. This is key to the simulation and formal verification approaches that will be carried out in this chapter.

Eliminating all ambiguity from the interpretation of the design diagrams (block instances and state machines) – on the one hand – and translating these diagrams into a formalism that is already equipped in terms of formal verification – on the other hand – requires us to enhance these SysML design diagrams with a formal semantic. We can

Chapter written by Ludovic APVRILLE and Pierre DE SAQUI-SANNES.

do this with the help of an Automated Verification of real time software (AVATAR) language [APV 11a], which we will discuss in this chapter. AVATAR is the language supported by TTool [APV 11b]. TTool is a free software that is interfaced with the UPPAAL tool [BEN 04] in order to verify the logic and temporality of the designs modeled in SysML.

This chapter revisits the SysML model of the pacemaker from Chapter 4 and further enhances it in order to perform a logical and temporal analysis. We will expand the original SysML requirements diagram and formally link the requirements to the properties that must be verified.

The temporal property expression language (TEPE) for expressing properties and their integration in the "model verification" method, which is supported by TTool, are of crucial importance to the presentation of the TTool/AVATAR environment described in section 5.2. This AVATAR model of the pacemaker has a "documentation" part that is formed of requirement diagrams and analysis diagrams (section 5.3) and a part deemed "executable" formed of architectural design diagrams (section 5.4) and behavior diagrams (section 5.5).

Design diagrams can be debugged with the simulator integrated in TTool. This debugging feature is presented in section 5.5, and illustrated via the most significant diagrams of the AVATAR model of the pacemaker.

Section 5.6 details the formal verification. From the various functioning modes of the pacemaker, this chapter retains the VVI mode for the conciseness of the models and the simulation traces. The formal verification of the same model showcases the interest in the "press button" approach, carried out by TTool without hiding the problem of the information feedback from UPPAAL toward TTool. The diagrams for expressing properties (TEPE language) allow us to describe the properties and translate them into observers that are useful for guiding the formal verification. The set forms a simulation environment and a formal verification environment for timed SysML models that section 5.7 compares against other tools and SysML approaches.

Section 5.8 summarizes the contributions of the chapter and opens up new perspectives on the transition between the SysML/AVATAR model and the AADL model of the same pacemaker.

5.2. The AVATAR language and the TTool tool

The AVATAR approach comes with a method (section 5.2.1), a SysML-based language (section 5.2.2), enhanced by TEPE (section 5.2.3) and supported by TTool (section 5.2.4).

5.2.1. *Method*

The SysML language standardized by the Object Management Group (OMG) is merely a notation, by no means a method. Initially, the AVATAR language was in the same situation. We have attached to it a method that is applicable to a large scale of systems. The seven stages are linked to one another in this particular order, although loopbacks may occur:

1) gathering the requirements;

2) expressing simplifying hypotheses;

3) carrying out an analysis guided by the use cases that have been documented by scenarios and activity concatenations;

4) designing the architecture and behaviors;

5) expressing the properties that need to be verified;

6) checking the design diagrams against the requirements and properties by combining simulation techniques and formal verification techniques;

7) expressing the allocations between the model elements and the building of traceability matrices.

A loopback can take place between any specified i stage toward a previous j stage with $j < i$, even if we will always prefer to go as far as possible – that is up to the final stage 7 – in each iteration of the method.

5.2.2. *AVATAR language and SysML standard*

The AVATAR language reuses the nine SysML diagrams, their syntax and their semantic insofar as the syntax is defined in a precise manner by the [OMG 11] standard. The AVATAR language is better equipped and associated with a method that caters to the needs of a larger class of real-time systems. This method guides the way we introduce the AVATAR diagrams and compare them with their SysML standardized version.

In the early stages of the method, requirements are captured in a tree-like structure. Each requirement of the tree contains attributes, as well as connections with other elements of the model. One important connection is a reference to properties that must be formally verified on the functional architecture.

The analysis phase is classically based on a use case diagram that is documented by sequence diagrams and activity diagrams. AVATAR reuses the interaction overview diagram (IOD) of Unified Modeling Language (UML) 2 to structure the scenarios that describe the different sequence diagrams. The latter support not only relative time and absolute time operators but also the manipulation of timers throughout the building operation, the timeout test and the reinitialization. Besides managing time operators, we must also represent the reactive character of real-time systems. To do this, the AVATAR sequence diagrams model the asynchronous interactions and the synchronous ones. Let us note, however, that continuous flows are not supported. This limitation is due to the fact that, nowadays, we do not yet have the simulation tools and formal verification tools capable of processing discrete and continuous flows from an AVATAR model.

This restriction can be found in the design diagrams that define the system entities in terms of architecture as well as in terms of behavior. In terms of architecture, AVATAR uses the same diagram for, on the one hand, representing the blocks (hence the typing of architectural elements) and, on the other hand, representing the block instances (that are, in the same diagram, composed in the sense of the composition between intercommunicating timed state machines). In terms of behavior, state machine diagrams support time intervals (clause "after[min, max]") as well as the operations on the already mentioned timers for the sequence diagrams. AVATAR here makes the hypothesis of a "closed" system, that is, the hypothesis that the signal exchange with the environment of the system must be explicitly modeled in the design.

This set of diagrams is a support for the method presented in the previous section, with the aim of validating the design architecture and the behavior automata. To this end, the block instance diagrams and the state machine diagrams are equipped with a formal semantic allowing their translation into timed automata accepted as input by UPPAAL tool. This adds more formalization to SysML.

5.2.3. *The TEPE language for expressing properties*

The formalizing of the AVATAR language and the gateway established between the TTool and UPPAAL tools partake in the implementation of a formal verification approach. This approach classically refers to the notion of a "property to be verified". Often derived from requirements, these properties are commonly expressed in a formalism other than SysML [FAR 06, LEB 10, SAT 10]. On the contrary, the introduction of properties in the SysML model of the system is an AVATAR specificity that relies on an extension of SysML parametric diagrams in order to define a graphic language for expressing properties: TEPE [KNO 11a].

The SysML language defines parametric diagrams to establish the relations between the attributes of the architecture blocks and, more generally, for defining equations that involve the elements of one or several diagrams of a SysML model. Therefore, we believe that these parametric diagrams provide an excellent framework for expressing the properties that must be verified. With TEPE, the properties are described, on the one hand, via logical and temporal relations between the occurrences of events, and on the other hand with the help of the relations between values of the block attributes. The TEPE language for expressing time properties is born out of these two types of relation that gain in terms of ease of use, without, however, losing in terms of their expressiveness. Last but not least, the TEPE formal semantic is based on the temporal types of logic: *metric temporal logic* (MTL) [KOY 92] and *fluent linear temporal logic* (FLTL) [LET 05].

5.2.4. *TTool*

The AVATAR language is fully supported by the free software tool TTool [APV 11b] developed for Linux, Windows and MacOS. The default installation of TTool equips us with diagram editors and a simulator. TTool implements gateways toward three tools that are developed by other laboratories: UPPAAL for the formal verification of the logical and temporal properties, Proverif for the security properties verification [ABA 02] and SocLib for the virtual prototyping of the software and hardware of real-time systems. The absence of encrypted communication in the pacemaker system and the delegation of software implementations to the Modeling and Analysis of Real-Time and Embedded Systems (MARTE) and Architecture Analysis & Design Language (AADL) approaches means we will use neither Proverif nor SocLib in the rest of this chapter. On the contrary, UPPAAL will be used in this chapter to verify the constraints of functioning safety.

5.3. An AVATAR expression of the SysML model of the enhanced pacemaker

The AVATAR modeling is presented in two stages: the functioning and hypotheses of the modeled system, and the representation of the requirements of said system.

5.3.1. *Functioning of the pacemaker and modeling hypotheses*

In making up for the insufficiency of a heart that no longer beats correctly, a well-functioning pacemaker can find itself in one of the following modes, which are characterized by three letters:

– the first letter (A, V or D) indicates which one is being electrically stimulated (or "paced") – the atrium, the ventricle, or both – and which must therefore receive an impulse;

– the second letter (A, V or D) indicates which room (ultimately both rooms) will be electrically monitored and thus observed;

– the third letter I, T or D identifies the modes *inhibited*, *self-triggering* and *dual pacing*, respectively. In mode I, a heart contraction detected as an event coming from the heart in the shared delay will stop the pacemaker from sending an electrical impulse. In mode T, a heart event will immediately stimulate the pacemaker. Thus, in VVI mode, the ventricle is stimulated if an event does not come from the ventricle. If an event does come from the ventricle then the stimulation is inhibited. This particular mode will be used in this chapter to illustrate the formal verification guided by several observers.

This chapter adds other simplifying objects to the model. We will assume, in particular, that the pacemaker was correctly implanted and that the patient is in non-ambulatory mode. The pacemaker is deemed to have been correctly initialized. Its software and hardware components do not fail. Maintenance, replacement, recycling of the pacemaker are not tackled. We extract a simplified architecture from this global architecture and use it as a support for a pedagogical presentation of the simulation modes and the formal verification modes supported by TTool.

5.3.2. Requirements diagram

The AVATAR requirements diagram in Figure 5.1 reviews the SysML requirement diagrams in Chapter 4. At the same time, the diagram emphasizes the main goals of the application, the modeling hypotheses, the refining relations between the requirements, and the composition relations. We have enhanced this diagram with links between requirements and "properties to be verified". For the time being, these properties are identified by a name. They will be detailed when the architectural elements such as the names of the signals and the block attributes have been defined.

In this chapter, we will focus on the properties related to the VVI mode. More precisely, the left-hand side vertical in Figure 5.1 is dedicated to requirements, starting from the most fundamental requirement (every cardiac problem must be under the control of the pacemaker) toward the most concrete ones. The right-hand side vertical completes the diagram with hypotheses regarding the patient and the pacemaker, most notably regarding the functioning of the system itself. At the bottom of the figure, the "verify" relations link the VVI stimulation requirement node to the properties that we will verify in section 5.6.

5.4. Architecture

The keystone of an AVATAR design is the block instances diagram, which reflects the architecture of the modeled system. The simulation and formal verification of an AVATAR model need to have a "closed" model. This is why the block instance diagram in Figure 5.2 describes not only the internal architecture of the pacemaker, but also the connections between the pacemaker and the heart that it observes and stimulates. Let us note that Figure 5.2 describes an organic architecture that, in the process of system engineering, comes *after* the functional architecture; this latter architecture has not been described in this chapter because it has already been described in Chapter 4.

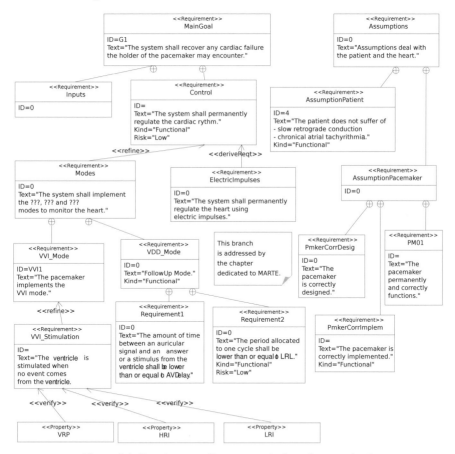

Figure 5.1. *Requirements diagram: goals, hypotheses, refined requirements and properties*

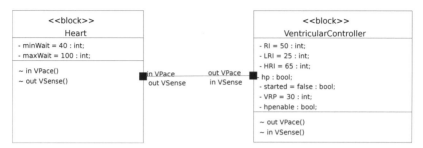

Figure 5.2. *Sample of the block instance diagram*

The AVATAR language groups together, in a single diagram, the definition diagrams of the blocks and of the internal view of the diagrams in order to form a block instance diagram. The connectors that link the ports of the two block instances enable us to define the synchronous communications between the local state machines and these two instances.

5.5. Behavior

All of the block instances represented in Figure 5.2 have a behavior that is described by state machines. We have chosen to present those of the controller of the ventricle (*VentricularController*) and of the hearts in Figures 5.3 and 5.4. The first diagram (Figure 5.3) emphasizes two main states (*WaitRI* and *WaitVRP*) each corresponding to the execution of an impact on the heart – if the heart has not carried out a normal activity – and waiting for the next cycle.

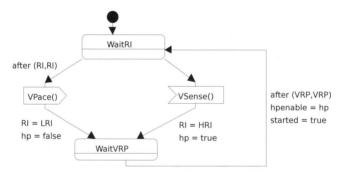

Figure 5.3. *State machine of the ventricle controller*

Given that block instances display behaviors, it is therefore possible to simulate them with TTool. This allows us to play out scenarios of the execution of the model, either step by step or in automatic mode. The parts of the model that are being explored can be quickly identified on the diagrams themselves and the simulation traces take on the shape of sequence diagrams, such as the one presented in Figure 5.5. These traces emphasize the synchronous and asynchronous communications between the instances (i.e. the changes in the value of the variables) and finally the delays between the actions or the communications. The simulator of TTool manages a single clock, global and shared by all the block instances that compose the architecture of the system. The time unit is global to the system and corresponds, for example, to 1 sec. The notation "@50" that appears on the simulation layout in Figure 5.5 expresses the absolute date that takes place 50 units after the start of the simulation. The numbers associated with the thin vertical triangles specify the progression of time.

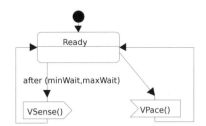

Figure 5.4. *State machine of the heart controller*

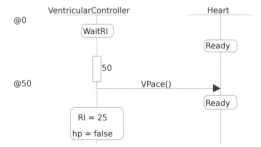

Figure 5.5. *Sample of the sequence diagram obtained from a simulation of the system model*

5.6. Formal verification of the VVI mode

The practice of formal verification allows us to distinguish between properties that are deemed as "general", which are applicable to a large class of systems (section 5.6.1) and the properties that are specific to the system currently studied (section 5.6.2).

5.6.1. *General properties*

As with the majority of real-time systems, the pacemaker must satisfy general properties such as the absence of a *deadlock*, the absence of an internal cycle without a *livelock* and the ability to return to the initial state.

The verification of these properties with the help of the UPPAAL tool needs the translation of the AVATAR model into timed automata. This latter task is entirely managed by TTool, which proposes, in native mode, to verify the absence of a deadlock. The screenshot in Figure 5.6 thus shows that the AVATAR model of the pacemaker does not have a *deadlock*. This property has been verified without having to write or read a line of the "timed automata code", which is automatically generated by TTool and injected in UPPAAL.

Figure 5.6. *Verification window: researching the* deadlock

5.6.2. *Expressing properties using TEPE*

We have modeled in TEPE the three properties defined during the stage of modeling the requirements (see section 5.3.2):

– LRI: when the *pace* on the hysteresis is deactivated, a *pace* on the ventricle or a *sense* action of the heart must be carried out before LRI time units;

– HRI: when the *pace* on the hysteresis is activated, a *pace* on the ventricle or a *sense* action of the heart must be carried out before HRI time units;

– VRP: a *sense* action cannot be carried out before VRP time units have gone by since the previous *sense* action or a *pace* action.

For example, Figure 5.7 represents the *LRI* property. The TC operator represents a time constraint of an LRI value. Thus, once the hysteresis is deactivated, an "Action" (that is either a *VPace* signal or a *VSense* signal, see the "*Alias*" operator) must be carried out within the system before LRI time units. Similarly, it is possible to easily model the *HRI* property graphically.

The PropVRP property is more difficult to model because it refers to more particular states of the controller's state machine. Thus, its modeling rests upon the use of entry signals in the states *WaitVRP* and *WaitRI* (see Figure 5.8). The property models two sub-properties that are put in relation with one another via the operator *OR*. A sub-property expresses the case where the system has not yet carried out a first cycle, that is, the value of *started* is *false*. The second sub-property is uniquely applied in the case where a first cycle was carried out: thus, the system has already reached the *WaitVRP* state, which translates into the signal *enterState__WaitVRP*). In this case, the system must wait for a *VRP* period before carrying out the following cycle whose start is translated by entering into the *WaitRI* state (TEPE signal *enterState__WaitRI*).

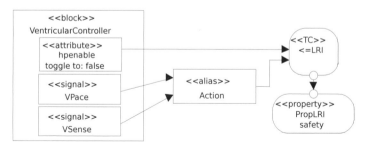

Figure 5.7. *TEPE modeling of a safety property: LRI*

The verification of the properties already described rests either on their expression in the form of a temporal logic formula (section 5.6.3) or by using observers (section 5.6.4).

5.6.3. *The use of temporal logic*

Verification often involves expressing the system in a formal language (for instance, language of temporal ordering specification (LOTOS)), expressing the properties that need to be verified in a temporal logic (for instance, Computational Tree Logic (CTL)), and then in giving the specification of the system and the properties to a *modelchecker* as input. The system is transformed into a system of labeled transitions where the properties are studied by the *modelchecker*. The most common kinds of temporal logic are linear temporal logic (LTL), CTL and PSL/Sugar.

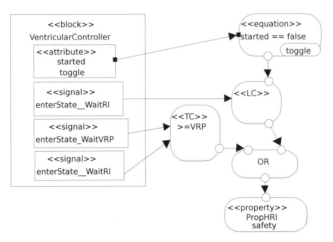

Figure 5.8. *TEPE modeling of a safety property: PropVRP*

With TTool, we can isolate a CTL formula that is analyzed by the UPPAAL *modelchecker*. Since this formula refers directly to the UPPAAL automaton, one needs to understand the UPPAAL automata generated by TTool in order to understand this formula. To show the difference between a property represented graphically using TEPE, and a corresponding CTL formula, here is the CTL formula of an LRI property, in UPPAAL format:

```
A[] VentricularController__0.hp  imply
VentricularController__0.h <= VentricularController__0.HRI
```

and that of the PropVRP property is:

```
A[] ((VentricularController__0.id0 && VentricularController__0.started) imply
VentricularController__0.h__ >= VentricularController__0.VRP
```

Thus, Figure 5.9 shows the formula composition window of a CTL formula and its verification after using TTool.

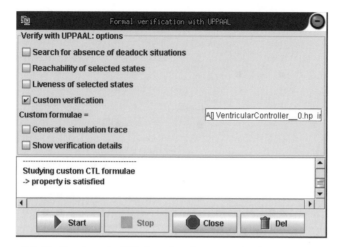

Figure 5.9. *Verification of a CTL formula after using TTool: verification of LRI*

5.6.4. *Observer-guided verification*

The verification guided by observers consists of adding to the design model new blocks that are in charge of executing one particular action (e.g. executing an "error" action) whenever the property they are observing is no longer satisfied. Observation is done by introspecting other model elements: attributes and signals.

– The observation blocks are hypothesized to be non-intrusive on the system observed, that is their addition to the model is to under no circumstances alter the result of the observed properties.

– The addition of observation blocks can sometimes lead to the alteration of the design in order to add signals emitted by the observed blocks. This signal addition is always done while respecting the non-invasive aspect of the observer, that is the observer must, all the time, be ready to accept the synchronous signals that are added to the system for observation.

For example, Figure 5.10 presents the block instance diagrams to which observers for the three properties studied in this chapter have been added. For each new block, we must add, of course, a state machine.

The state machine of the observer that studies the LRI and HRI properties is represented in Figure 5.11. The signals sent by the controller of the ventricle are analyzed in order to determine if a clearly specified time interval flows between receptions or not. This time interval is modeled by the starting, the timeout and the stopping of a timer that serves as a stopwatch.

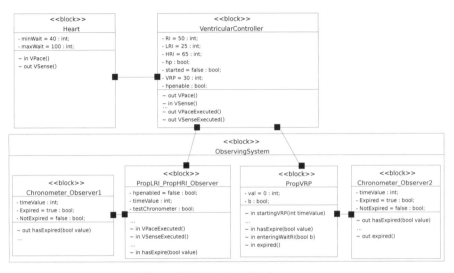

Figure 5.10. *Design with observers*

TTool allows the automatic research of the reachability of states or actions of state machines. The verification methodology that makes use of observers is based on a test for the observers' non-reachability of error states. This test is automatically carried out by TTool, once more relying on the automatic transformation of the model and the observers into timed automata, then using the UPPAAL modelchecker with CTL formulas obtained by analyzing the SysML modeling and its transformation into timed automata. Figure 5.11 shows the result of this verification methodology applied to the pacemaker, more particularly the fact that the error state of the observer of LRI and HRI properties is not reachable. With recent computers, this verification result is obtained in a few seconds: this can, of course, be explained by the fact that we have only considered a subset of the model.

The observers have been manually added to the design. At the time of writing this chapter, the TEPE properties cannot be automatically transformed into observers. We are currently working on enhancing TTool with this automatic generation, as is the case for other profiles: Timed UML and RT-LOTOS Environment (TURTLE) [FON 08] and Design Space Exloration Based on Formal Description Techniques, UML and Systemc (DIPLODOCUS) [KNO 11b].

5.6.5. *Coming back to the model*

When we carry out simulations or verifications of a model, there arises the question of the interpretation of the results issued from said simulations and verifications in relation to the model provided at the entry of the simulator or the verifier.

TTool simulations of AVATAR modelings are carried out directly on the diagrams. In particular, TTool triggers the state machines of the simulated system directly and produces in return a sequence diagram whose names of entities and actions are exactly those of the model.

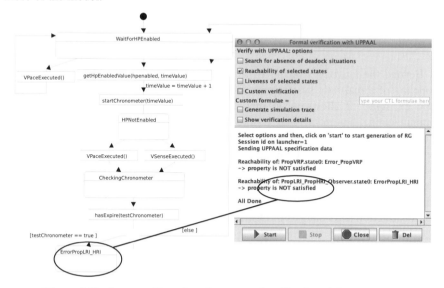

Figure 5.11. *State machine of an observer and verification of the error state non-reachability of that observer*

In the case of verifications, the tests are carried out on a model in the shape of timed automata, a model that is structurally different from the corresponding AVATAR model. In the case of a non-verified property, the UPPAAL modelchecker shows a trace leading to the violation of the property. At the time of writing this chapter, this simulation trace must be analyzed directly in UPPAAL, since TTool is not able to transform it in one AVATAR trace. The solution for this problem will be the focus of our future work.

5.7. Related work

Our contribution consists of a modeling environment (AVATAR) and a tool that supports this language (TTool). In this section, we compare these two contributions with similar ones.

5.7.1. *Languages*

Within the theoretical approaches that try to build bridges between the SysML language and the formal methods supported by model verification tools, the AVATAR language specifies the semantic of the SysML model diagrams and integrates the expression of properties straight into a SysML diagram, where other environments usually delegate this modeling function to an external formalism, for example to temporal logic. As opposed to the Clock Constraint Specification Language (CCSL) language (associated with the UML-MARTE profile), the TEPE language for expressing properties is supposed to be more accessible to people that are usually put off by formal languages.

Beyond the languages of the UML family, many other environments allow the modeling of a real-time system and its formal verification. UPPAAL [BEN 04] is a good example: it allows us to model a system in the shape of a set of timed communicating automata. The formal verification of CTL properties is directly integrated to the environment bearing the same name (UPPAAL). On the other hand, the language in the shape of automata barely supports a system engineering process, and is not commonly used for documentation purposes. This observation goes for other formalisms as well, such as the Petri nets supported by the TIme Petri Net Analyzer (TINA) tool [BER 06] or even the LOTOS process algebra supported by the Construction and Analysis of Distributed Processes (CADP) tool [GAR 07].

5.7.2. *Tools*

UML/SysML toward Electronic System Level (ESL) [VIE 06] is a modeling environment for systems-on-chip, providing simulation and formal verification from the models made up of UML sequence diagrams. The simulation and verification results particularly allow us to obtain the worst case execution times. But even there, the engineering phase is not wholly covered. In addition, the environment is very limited in terms of communications.

The SysML-based environment presented in [DA 09] facilitates the verification – using UPPAAL – of logical and temporal properties on the models produced with the help of the Rhapsody tool. This approach does not support the stage of requirement identification. In the absence of high-level language for expressing the properties, we must use the CTL formulas. TTool allows us, on the contrary, to verify the accessibility and the liveness of the actions by adding, with a simple mouse-click, markers to the respective actions.

The OMEGA environment [OBE 09] supports the phase of capturing the requirements because of its SysML support, but does not allow us to integrate these requirements (and the properties) into the formal verification approach. OMEGA and TTool do not support the continuous flows of SysML, as opposed to the Artisan tool, which supports, even more, the probabilities on the transitions, as well as the interruptible areas.

Developed in the Eclipse environment that has recently been enhanced with the Papyrus editor, the TopCased toolset [FAR 06] brings together model analysis tools and code generation tools that exploit the "model transformation" approach. TTool relies on a less flexible, but formalized environment. This remark applies both to system design and properties capture.

The research into new electronics design techniques has also stimulated research connecting SysML to formal methods. Thus, SysML models can be translated into Very-high-speed integrated circuits Hardware Description Language – Analog and Mixed-Signal (VHDL-AMS) [REA 11] or into *simulink* [VAN 06]. In mechanical engineering, SysML is also used in conjunction with other specific languages such as Modelica [PAR 10]. This language describes mechanical, electrical, hydraulic and thermal components of a system in the shape of a set of equations that a simulator can resolve at each clock cycle.

5.8. Conclusion

A SysML model cannot stop at the stage of engineering drawing. A SysML model analysis tool that brings together simulation and formal verification detects design errors earlier in the lifecycle of the system. This type of tool confronts a SysML model with the requirements and properties that are suited for working with extended versions of the SysML language. The TTool used in this chapter is an example of combining a pacemaker model expressed in the SysML/AVATAR language with safety properties.

Starting from the SysML model proposed in Chapter 4, we have built an AVATAR model oriented toward the analysis of the logical and temporal behaviors of the controller of the pacemaker. The study of the VVI mode has allowed us to illustrate the use of AVATAR and TTool throughout this chapter. We have successively presented the debugging simulation, the model control verification and verification guided by observers.

These three complementary modes of analysis have allowed us to formally test the satisfaction of several requirements. These results of the simulation and verification rest on the model provided by the user in TTool, on the absence of errors in the process of translation into timed automata and on the absence of errors in the UPPAAL *modelchecker*. The use of a verification tools translator, where the tools are certified or tested, would allow us to ground the results of our verification more firmly.

5.9. Appendix: TTool

TTool can be downloaded from ttool.telecom-paristech.fr. It runs in Windows, MacOS and Linux. Once we have installed the tool, we need to double-click on *ttool.jar*.

TTool incorporates a graphic editor, a model simulator and code generators that are dedicated to external tools. The first click gives us access to the editor and the simulator. If the external tools have been uploaded and the configuration file has been adjusted accordingly, then the formal verification can be managed from the TTool menus, by referring to the identifiers of the AVATAR model (and not to the code generated from the AVATAR language). The website dedicated to the tool and to the AVATAR profile offers a tutorial for using AVATAR, examples of models and presents the AVATAR language and the available verification techniques.

The diagram editor supports AVATAR, a version of SysML enhanced with real-time extensions and a formal semantic. The simulator that is incorporated in TTool facilitates the development and the debugging of the models. The interfaces of the formal verification tools UPPAAL and ProVerif authorize more in-depth model analyses without requiring that the formal methods be mastered. These two external tools, respectively, target the properties of safety and security.

5.10. Bibliography

[ABA 02] ABADI M., BLANCHET B., "Analyzing security protocols with secrecy types and logic programs", *29th Annual ACM SIGPLAN - SIGACT Symposium on Principles of Programming Languages (POPL 2002)*, ACM Press, Portland, Oregon, pp. 33–44, January 2002.

[APV 11a] APVRILLE L., SAQUI-SANNES P.D., "AVATAR/TTool: un environnement en mode libre pour SysML temps réel", *Génie Logiciel*, vol. 98, pp. 22–26, September 2011.

[APV 11b] APVRILLE L., Webpage of TTool, 2011. Available at ttool.telecom-paristech.fr/.

[BEN 04] BENGTSSON J., YI W., "Timed automata: semantics, algorithms and tools", in REISIG W., ROZENBERG G. (eds) *Lecture Notes on Concurrency and Petri Nets*, LNCS 3098, Springer-Verlag, pp. 87–124, 2004.

[BER 06] BERTHOMIEU B., VERNADAT F., "Time petri nets analysis with TINA", *3rd International IEEE Conference on the Quantitative Evaluation of Systems (QEST 2006)*, Washington, DC, USA, pp. 123–124, 2006.

[DA 09] DA SILVA E.C., VILLANI E., "Integrating SysML and model-checking techniques for the v and v of space-based embedded critical software", *Brasilian Symposium on Aerospace Engineering and Applications*, São José dos Campos, Brazil, 2009 (in Portugese).

[FAR 06] FARAIL P., GAUFILLET P., CANAL A., *et al.* "The topcased project: a toolkit in open source for critical aeronautic systems design", *Embedded Real Time Software (ERTS '06)*, Toulouse, France, November 2006.

[FON 08] FONTAN B., Méthodologie de conception de systèmes temps réel et distribués en contexte UML/SysML, Doctoral Thesis, Laboratory for Analysis and Architecture Systems (LAAS-CNRS), 2008.

[GAR 07] GARAVEL H., LANG F., MATEESCU R. *et al.* "CADP 2006: a toolbox for the construction and analysis of distributed processes", *Computer Aided Verification (CAV'2007)*, Berlin, Germany, vol. 4590, pp. 158–163, 2007.

[KNO 11a] KNORRECK D., APVRILLE L., DE SAQUI-SANNES P., "TEPE: a SysML language for time-constrained property modeling and formal verification", *ACM SIGSOFT Software Engineering Notes*, vol. 36/1, pp. 1–8, January 2011.

[KNO 11b] KNORRECK D., UML-based design space exploration, fast simulation and static analysis, PhD Thesis, EDITE, Ecole Doctorale Informatique, Télécommunications et Electronique, Telecom ParisTech, 2011.

[KOY 92] KOYMANS R., *Specifying Message Passing and Time-Critical Systems with Temporal Logic*, Springer-Verlag New York Inc., Secaucus, NJ, 1992.

[LEB 10] LEBLANC P., Modélisation SysML exécutable d'un système événementiel et continu, IBM White Paper, 2010. Available at files.me.com/jmbruel/gblurt.

[LET 05] LETIER E., KRAMER J., MAGEE J., *et al.* "Fluent temporal logic for discrete-time event-based models", *Proceedings of the 10th European Software Engineering Conference*, ESEC/FSE-13, ACM, New York, NY, pp. 70–79, 2005.

[OBE 09] OBER I., DRAGOMIR I., "OMEGA2: a new version of the profile and the tools", *14th IEEE International Conference on Engineering of Complex Computer Systems, UML-AADL'2009*, Potsdam, pp. 373–378, 2009.

[OMG 11] OMG, SysML, 2011, Available at www.sysml.org/.

[PAR 10] PAREDIS C.J., BERNARD Y., BURKHART R.M., *et al.* "An overview of the SysML-modelica transformation specification", *International Council on System Engineering INCOSE '2010*, 2010.

[REA 11] REALTIME-AT-WORK, SysML-companion: virtual prototyping from SysML models, 2011. Available at www.realtimeatwork.com/software/sysml-companion/.

[SAT 10] SATURN P.F., SATURN, SysML bAsed modeling, architecTUre exploRation, simulation and syNthesis for complex embedded systems, 2010. Available at www.saturn-fp7.eu/.

[VAN 06] VANDERPERREN Y., DEHAENE W., "From UML/SysML to Matlab/Simulink: current state and future perspectives", *Proceedings of the Conference on Design, Automation and Test in Europe (DATE '06)*, European Design and Automation Association, Leuven, Belgium, pp. 93–93, 2006.

[VIE 06] VIELHL A., SCHONWALD T., BRINGMANN O., *et al.* "Formal performance analysis and simulation of UML/SysML models for ESL design", *Proceedings of the Conference on Design, Automation and Test in Europe (DATE '06)*, Munich, Germany, pp. 1–6, 2006.

PART 3

MARTE

Chapter 6

An Introduction to MARTE Concepts

6.1. Introduction

As part of its activities in the field of model engineering, the Object Management Group (OMG) has standardized an extension of the unified modeling language (UML) for the domain of real-time and embedded software systems; this extension is called Modeling and Analysis of Real-Time Embedded Systems (MARTE). The object of this chapter is to broadly sketch out this norm in order to provide the information needed for exploring it further. Furthermore, we will review certain specific points in more detail, such as the support MARTE offers for a component-based modeling of an application and, in the context given, we will explain how MARTE enables us to model such properties as concurrency or time. We will also look at another important aspect of the real-time embedded field, that is platform modeling, before we move on to an overview of the MARTE facilities with regard to the quantitative analysis of the models, more particularly, the scheduling analysis.

However, this section does not aim to be a user guide for MARTE. Rather the objective is to provide the readers with a minimum of information necessary for providing a solid introduction in the analysis and the use of this norm.

6.2. General remarks

In September 2003, the OMG published its first modeling norm in the real-time domain: the profile named *UML profile for schedulability, performance and time* (SPT) [OMG 05a]. The application field of SPT was restricted to the quantitative

Chapter written by Sébastien GÉRARD and François TERRIER.

analysis of UML models. It particularly enabled us to annotate the UML models for carrying out scheduling analyses that are part of the set of rate monotonic analysis (RMA) techniques, as well as performance analyses carried out using techniques based on queuing theories.

Shortly after it came out, both manufacturers and researchers published works on this norm, particularly discussing how adequate it was for industrial and academic practices within the field of model-based engineering, in particular in the field of development via embedded real-time system models [GER 03, CEA 05]. Among other things, we can mention the CEA, the INRIA and Thales that, throughout several projects carried out within a common research programme, CARROLL, have been the main actors in the definition and the publication of a new norm, MARTE. The latter is destined to replace SPT and cover a larger application domain: the field of development via real-time embedded system models. More specifically, it had to cover specific needs in terms of software modeling and hardware platforms, and to support component-based modelings [OMG 05b].

The first version of the MARTE norm was published by OMG in November 2009 (the current version at that point was version 1.2). Let us note that all the norms published by the OMG are public and available for free on the OMG website[1].

Let us also note that all of the norms published by the OMG follow a maintenance process defined in [OMG 09a]. Among others, each norm is updated in order to correct the problems raised by its users and implementers throughout revision groups called *Revision Task Force* (RTF). MARTE is well into its third revision cycle, which should finish in 2013 and give way to version 1.3.

6.2.1. *Possible uses of MARTE*

Clause 6.2.3 of the MARTE norm defines a certain number of expected use cases, including underlying actors. Here, we find two main categories of norm users: the methodologists and the providers of the execution infrastructure on the one hand, and the language users on the other hand.

The first category corresponds to those who, relying on MARTE, will define a use for it that corresponds to the requirements of a specific domain, will provide support for the execution of resulting models or will even further specialize or extend a part of it regarding a particular technology or a particular field.

The second category designates those who will implement the results of the work carried out by the participants in the first category.

---

1 See www.omg.org/spec/.

REMARK 6.1.– It is worth noting that, at the beginnings of UML, MARTE was a modeling language, and as such, its specification does not indicate a particular use in detail. It is the responsibility of the members in the first use category to provide this information to engineers, who are the users of this norm.

The chapters that follow in this part illustrate certain uses of the norm. Chapter 7 shows how MARTE can be used to model the application of the pacemaker, and Chapter 8 showcases the validation possibilities using this same model. Finally, Chapter 9 shows the component-oriented design starting from a MARTE-like model and the process of code generation.

6.2.2. How should we read the norm?

The field covered by the MARTE norm is vast, since it must cater for all the needs related to modeling of the design and validation of embedded real-time systems, and must do this by taking into account the greatest possible number of technologies. For this reason, the MARTE specification document is almost 800 pages long. To make its use more practical, its design was conceived to be modular.

The norm contains the definition of conformity cases (see clause 2 of [OMG 09b]) by facilitating a use that is both bespoke and optimal. The definition of these conformity cases rests on a set of use cases that are expected of the norm. The conformity cases are specified on two levels of abstraction: basic and complete. Each level specifies a concrete set of extension units considered to be mandatory for the level considered. The basic level is a subset of the complete level and the extension units of the complete level are considered to be optional at the basic level.

MARTE has taken advantage of the possibility to decompose a profile into a hierarchy of subprofiles. The advantage of this type of structure is to separate the preoccupations from the profile thus facilitating its comprehension. Another advantage is that each subprofile can be used independently (or, potentially, along with the subprofiles it depends on). The extension units of the conformity cases thus reflect the architecture of the MARTE profile and correspond to its different subprofiles.

As for example, we can detail the conformity case « software modeling ». Its subject is to offer support for modeling of the software aspects of real-time embedded applications, including taking into account non-functional properties (NFP). In order to reach this goal, it needs the following MARTE subcomponents:

– at the basic level: « generic resource modeling », « NFP », « Time » and « high-level application modeling (HLAM) »;

– at the complete level: « generic resource modeling », « NFP », « value specification languagey (VSL) », « Time », « Clock Handling Facilities », « software resource modeling (SRM) », « generic component model (GCM) » and « HLAM ».

For all the details on the content of each conformity case, we welcome the readers to refer to Table 2.2 of the clause 2.4.2 of the norm [OMG 09b].

With the aim of facilitating its use, the organization of the chapters describing each of the MARTE subprofiles abides by the following pattern:

1) a first section delineating the contours of the subprofile concerned and its architecture;

2) a second section describing the model of the domain relative to the concern of the subprofile, which is accompanied by a general text broadly sketching the model. This section can itself be divided down into subsections depending on the size of the field concerned. The semantic description of each element is recorded in annex F of the norm. It is not necessary to read this annex at first glance in order to come back to the enumeration above when one must learn in more detail about the use of a subprofile;

3) a third section that is itself generally decomposed into two subsections:

    i) a first subsection that contains the set of UML profile diagrams describing the extensions defined in the subprofile,

    ii) a second subsection describing in detail each stereotype of the profile, including the rules for the well formation of the models relative to the use of the stereotype as well as possible extensions of the UML notation;

4) a fourth section that proposes a set of examples illustrating the previously defined extensions.

To conclude on this subject, let us mention that one of the reasons behind the choices made with regard to the definition of MARTE concepts was to minimize implementation efforts for the users. However, achieving this objective, especially providing adequate and precise capabilities for modeling in the real-time embedded field needs an additional activity regarding a more common, even if less constrained use of the UML.

6.2.3. *The MARTE architecture*

MARTE is a UML profile meant to support an engineering approach guided by the models for the development of embedded real-time systems. It is made up of a set of extensions that have been appropriated by general UML concepts, providing designers with first-class language constructions for their application domain.

Many of these extensions regard what we call the non-functional aspects of the embedded real-time systems. These concerns can be classified into two categories dealing with quantitative and qualitative aspects. However, they can be used to different levels of abstraction and can be applied to the modeling of applications, as well as to their analysis via the models (or to both at the same time). To satisfy all of these requirements, MARTE is structured into one hierarchy of subprofiles, as shown by the diagram in Figure 6.1.

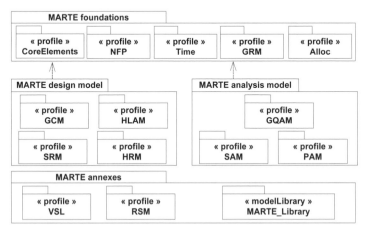

Figure 6.1. *View of the global MARTE architecture*

The "foundations" package (found at the top of Figure 6.1) defines the basic MARTE concepts and is made up of the following subprofiles:

– Core Elements: this subprofile mainly proposes concepts that allow us to model the operational modes of a system;

– NFP and VSL: the NFP subprofile provides the necessary modeling constraints for declaring, qualifying and applying non-functional information to a UML model. VSL is the indispensable companion of NFP since it enables us to specify the values of the non-functional annotations defined in the NFP subprofile;

– *Time* and *clock-constraint specification language* (CCSL): the *Time* subprofile defines the concept of time, which is of top priority for the embedded real-time systems. MARTE proposes three time models: a chronometric time model and a time model corresponding to the development approaches relying on the paradigm of the synchronous programming. On the other hand, this MARTE subprofile defines a set of low-level mechanisms for manipulating time, such as clocks, monitors or even timed events. CCSL is a textual language complementary to the *Time* profile allowing us to describe the constraints between the clocks. *Time* and CCSL are not detailed here; the readers wishing to know more on the subject should refer to [ESP 09];

– GRM: this subprofile satisfies a requirement that is important for the field of real-time embedded systems, namely platform modeling. The concept of platform has been abstracted from the norm, representing a computing resource, for instance. The GRM profile thus contains a taxonomy of resources allowing us to model all of the aspects of a platform at a level of system abstraction, that is regardless of software and/or hardware concerns. These two aspects are tackled in two other specialized MARTE subprofiles, SRM and *hardware ressource modeling* (HRM) subprofiles. Besides the support for platform modeling at the level of the system, GRM equally allows us to model the use of the platform and platform elements;

– Alloc (*Allocation Modeling*): the development cycles currently used in the field of real-time systems relies on the explicit modeling of an allocation model allowing us to connect the elements of the application model with the elements of the platform model. The "Alloc" profile defines the concepts necessary for the explicit description of this allocation model.

The profiles that have been described for MARTEs conceptual framework. Other, more complex concepts are defined more specifically on the basis of MARTE first in order to model real-time embedded applications in more detail, and second, in order to consider approaches that implement quantitative analyses of the models.

The development of model-based real-time embedded systems is mainly supported in a declarative manner by MARTE, namely on the basis of the annotations associated to the elements of a model, thus specifying their characteristics. The four MARTE subprofiles dedicated to this topic are as follows:

– HLAM: this submodel provides a set of extensions to the UML allowing us to annotate the elements of a model in order to define the real-time and embedded properties of a system. We can thus, for example, model the concurrence or specify a particular computing model.

– GCM: the component paradigm is essential for modeling embedded real-time systems and, in particular, their architecture. To do so, GCM reviews the composite structures of UML and extends one of its submodels in order to better address the domain-specific requirements in terms of component-based modeling. The main particularities of MARTE are specifying its semantics, particularly the link between the structural elements and the behavioral elements. MARTE clearly answers the question: what happens when a component receives a stimulus in one of its ports? On the other hand, the component model as defined in MARTE support the client–server model, but also introduces the communication via data exchange between components.

– SRM and HRM: these two subprofiles specialize the GRM subprofile in order to provide a basis for the modeling of software and hardware platforms. On the basis of

SRM, MARTE natively integrates the OSEK, ARINC and (partially) POSIX models, the three particular types of platforms.

Let us note the existence of another MARTE subprofile that is more particularly dedicated to the modeling of systems-on-chip: the *repetitive structure modeling* (RSM) subprofile. This allows us to describe systems-on-chip of the massively parallel computing type, such as systems that implement image processing algorithms. Thus, RSM allows us to describe this type of system and consider their generally regular structure, composed of various elements, by providing the concepts that favor a compact description of the models.

Finally, on the basis of the principles that are established in SPT, MARTE provides an extended support for the quantitative analysis approaches via real-time embedded system models, in particular the scheduling analyses and the performance analyses. The major addition in MARTE is the definition of a generic analysis subprofile defined as the "foundation subprofile" for the model-based quantitative analysis. The specter of the analyses allowed by MARTE has been expanded for scheduling and performance analyses.

6.2.4. *MARTE and SysML*

SysML and MARTE are modeling languages built on the basis of UML profiles. As a result, they can apply to the same model, making common use of these two dedicated languages possible. We must note that these two norms have been specified in parallel without any synchronization efforts. However, because of their specific focus, the engineering of embedded real-time models and the model based system engineering, these two norms are complementary on so many aspects. The capacities in terms of modeling of these two norms cover a large perimeter allowing us to satisfy numerous methodological approaches. However, while the combined use of these two norms is useful in numerous situations, the difficulty remains in making the best choices when combining the MARTE and SysML concepts. This task is definitely of the competence of previously mentioned methodologists, who must define an architectural framework, i.e. who must make the best use of the two norms according to the constraints of the respective domain and its preoccupations.

The only objective of this chapter is to talk about defining what constitutes a good use of MARTE and SysML. For a deeper study on the matter, the readers can refer to [ESP 09] who defines a certain number of use cases that brought the two norms together, norms that, conjointly, set the grounds for redefining new approaches and new uses.

6.2.5. *An open source support*

As part of their work on the Papyrus[2] *open source* tool that is dedicated to UML2 modeling, the CEA has developed an *open source* implementation of MARTE. This module, additional to Papyrus, integrates the set of extensions defined in the norm, the set of the libraries and the models specified in its annexes. It also proposes an advanced textual editor of the value description language, VSL.

6.3. Several MARTE details

The rest of this chapter details several MARTE aspects. We will not do a systematic review of all the concepts, but what we are concerned with here is presenting several concepts that are key for giving readers the basis for a first time use of MARTE.

6.3.1. *Modeling non-functional properties*

Let us recall that the main MARTE objective is to define a general framework allowing us to annotate the UML models in order to add to the functional model of the application the description of the so-called NFP, such as real-time properties or embedded properties. Answering this question in fact requires us to pay attention to the following corollary questions:

– how the NFP must be described, and in particular what are the different categories that must be considered?

– how should we associate NFP to a UML model?

– what are the possible relations between NFP and how should we describe and/or constrain them?

– what level of abstraction must we reach to facilitate the useability of this general framework?

– how should we expand the normalized framework, which is bound to be incomplete, so that we cover new fields or technologies?

– how should we formally describe the values of the NFP?

Let us note that the term "non-functional property" includes the terms "service quality" or even "extra functional properties", which we can encounter in other domains (e.g., in telecommunications).

Let us remember that the framework proposed by MARTE for solving the previous questions is made up of a dedicated profile, NFP, of a textual language, VSL, and of

---

2 See www.eclipse.org/papyrus.

several model libraries dedicated and predefined. NFP defines the concepts necessary for the description and the use of NFP. VSL allows us to describe precisely the values of the NFP, as well as certain qualifiers such as the unity of a value or its precision. VSL will only be approached in this chapter through examples. For more details on its semantics and its syntax, we welcome the readers to refer directly to annex B of the MARTE norm. Finally, the model libraries propose a set of predefined NFP and a certain number of model elements considered as MARTE utilities.

NFP defines a minimal set of concepts that are necessary for implementing the declaration of the NFP and the specification of the relations that can exist between these NFP. This set contains five concepts that we will detail now.

When we wish to give sense to these values, it is important to be able to specify the unity used for expressing the values. For in so doing, MARTE proposes to define these dimensions and their respective units. Specifying a dimension will consist in modeling a UML enumerated list annotated with the stereotype « Dimension » and its literal ... will be annotated with « unit ». The main property for « Dimension » is « symbol ». It is a chain of characters that denotes the symbol representing the dimension, for example « T » for time. The properties « unit » are baseDimension and offsetValue. These two properties allow us to specify for each unity a multiplying conversion factor and an offset value in relation to a particular unit of the dimension defined as the basic unit. Figure 6.2 hereafter presents an example from a MARTE library modeling the time dimension and its units (e.g. day, hour and minimum).

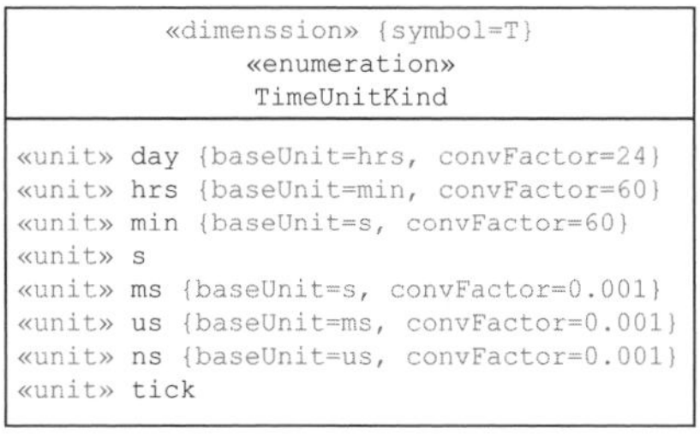

```
        «dimenssion» {symbol=T}
            «enumeration»
             TimeUnitKind

«unit» day {baseUnit=hrs, convFactor=24}
«unit» hrs {baseUnit=min, convFactor=60}
«unit» min {baseUnit=s, convFactor=60}
«unit» s
«unit» ms {baseUnit=s, convFactor=0.001}
«unit» us {baseUnit=ms, convFactor=0.001}
«unit» ns {baseUnit=us, convFactor=0.001}
«unit» tick
```

Figure 6.2. *Example of a definition of a dimension and its units: TimeUnitKind*

The first two concepts of a modeling subprofile and its NFP are the « Nfp » and « NfpType » concepts.

« Nfp » is used for explicitly marking a property in a model as representative for a non-functional information.

Figure 6.3 illustrates the example of the definition of a class modeling a processor. This processor has an identifier (id) and a (name) and two NFP allowing us to define the speed of the processor and its (powerConsumption).

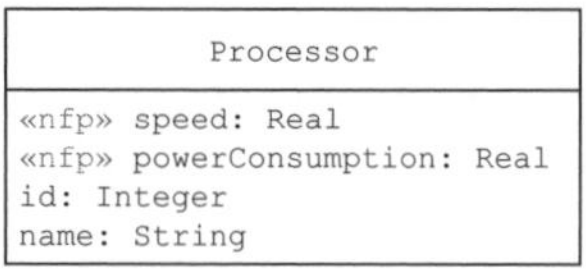

```
┌─────────────────────────────────────────┐
│                Processor                 │
├─────────────────────────────────────────┤
│ «nfp» speed: Real                        │
│ «nfp» powerConsumption: Real             │
│ id: Integer                              │
│ name: String                            │
└─────────────────────────────────────────┘
```

Figure 6.3. *Example of a use of the « nfp » concept*

« NfpType » is used for declaring types of non-functional property. This stereotype is an extension of the concept of UML « DataType ». What differentiates a UML data type from a UML class is that an instance of a datatype is uniquely identified through its value.

Before going any further in the description of the concepts of the NFP subprofile, let us make a short digression on the concept of stereotype and more precisely on the design styles of the stereotypes.

In the UML profiles, we can encounter two types of stereotype (the two forms are not mutually exclusive): on the one hand, the common use of the stereotypes consists of using them for modeling in a declarative fashion the metadata, thus augmenting the semantic of a UML mode. For example, we can annotate an operation of a class in order to specify its execution time or its memory size. In this case, the typifying of the properties of a stereotype consists of a primitive type, an enumerated type, a structured data type or a class.

The main limitation of this first method comes from the fact that the types of the stereotypical properties must be known in advance. Thus, the types used must be preemptively defined in the profile or in a model library that is either in the possession of the profile or imported by the profile. If, for one reason or another, a user wishes to modify an existing property in order to adapt its type to specific domain/requirement, the only solution is to redefine the profile and to introduce new characteristics or to modify the existing characteristics of the stereotypes.

To allow for a more flexible definition of the stereotypes and thus of a profile, one possible solution is to define the characteristics of the stereotypes via reference. To do this, we must typify the characteristics of a stereotype by using a UML 2 metaclass. By applying the stereotype to an element of the user model, we can give the semantic to the properties of the element other than its name or particular naming conventions. The models thus annotated are more easily exploited by external tools.

In the field of real-time embedded systems, as in the majority of specific domains, it is very difficult to define a complete taxonomy of useful types of non-functional property. To be sufficiently flexible and to thus allow considering non-identified or new requirements, the « NfpType » properties are defined via a reference as described in the previous section. The properties « NfpType », « valueAttrib », « unitAttrib » and « valueExpr » detailed below are thus typified by the « Property » UML metaclass:

– valueAttrib references the property of the annotated element, which will contain the value of a non-functional property;

– unitAttrib references the property of the annotated element that defines the dimension (i.e. the definition of the possible units) of the value of the non-functional property;

– valueExpr references the property of the element that specifies an expression (e.g. algebraic and logic). If such an expression is specified, this means that the value carried by the property referenced via valueAttrib is calculated by means of that expression. The VSL language is one possibility for describing this expression. This property aims to explicitly store in the model the way in which a derived value must be calculated. It is worth noting that the expression must specify a return type that is in accordance with the type of the property referenced by valueAttrib.

To complete its support of the non-functional properties, MARTE defines a model library containing a set of primitive types and structured types in a normative annex, predefining some NFP and a set of dimensions, including the description of the units for each of them:

– the library of MARTE primitive types, called « MARTE_PrimitiveTypes », takes the UML primitive types and adds the « Real » primitive type, and the definition of the usual mathematical operators needed for manipulating said types. On the other hand, MARTE also defines in its « MARTE_DataTypes » library, structured data types on top of UML data types. We may thus describe vectors, intervals and matrices of primitive types (e.g. integer vectors or real number matrices);

– the « MeasurementUnits » library predefines a set of dimensions and their respective units. Here, for example, the definition of units are relative to the time dimension (Figure 6.2) or even those relative to the size of computational data (bit, Byte, KB, etc.);

– the « BasicNFP_Types » library predefines non-functional property types such as « NFP_Duration ». This type allows us to define properties whose values express periods of time. In particular, the properties typified by « NFP_Duration » have a nominal value and can equally have a worst-case value (called « worst »), a value corresponding to the best case (called « best ») and a (« precision »). The precision is the same for all of the property values.

Finally, let us note that « MARTE_PrimitiveTypes » also has a particular type defined as an abstract type and used as a « mother-type » for other types of NFP – the NFP_CommonType (Figure 6.4).

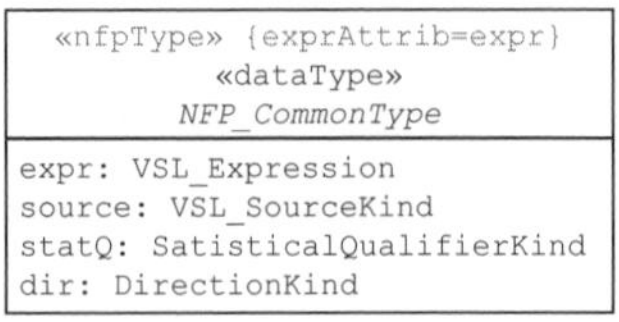

```
«nfpType» {exprAttrib=expr}
        «dataType»
       NFP_CommonType
------------------------------------
expr: VSL_Expression
source: VSL_SourceKind
statQ: SatisticalQualifierKind
dir: DirectionKind
```

Figure 6.4. *Details of the structured MARTE*
type::BasicNFP_Types::NFP_CommonType

The properties of the « NFP_CommonType » are the following:

– « expr » is a property that specifies the expression used, as the case may be, for calculating the value of the property. The type of this property is « VSL_Expression », which consequently imposes the use of VSL for describing the expressions for the derivation of the non-functional property values. This rule goes for all the other types defined in the « MARTE::BasicNFP_Types » library, since they all generalize the abstract type « NFP_CommonType »;

– « source » is a property that indicates the origin of the value indicated by the property. It can be estimated, measured, calculated or required (« source » has the value: *est*, *meas*, *calc* or *req*);

– « statQ » is a property that states the statistic nature of a measured value. The possible values for this property are defined by the enumerated type « SourceKind » whose possible literal values are: *max*, *min*, *mean*, *variance*, *range*, *percent*, *distrib*, *determ* or *other*.

Having specified dimensions, unities, NFP and non-functional property types, the « NFP » subprofile proposes the concept of « NfpConstraint » for describing constraints that have a bearing on NFP. We may distinguish between three types of constraints via the « kind » property; the possible values for the « kind » property are the following:

– *required* that indicates that the constraint specifies a minimum level required;

– *offered* indicates that the constraint specifies a space of values supported by the constrained elements;

– *contract* indicates that the underlying constraint specifies a relation between the values of the NFP, both required and provided.

A constraint has a context that identifies the element of the model, which determines its assessment context, and an ordered list of references toward model elements, which are in turn identified by the Boolean expression that specifies it.

Figure 6.5 shows two examples of constraints that can be applied to non-functional properties. At the top of the figure, we find the class called « Controller » that has four non-functional properties, « procUtiliz, « schedUtiliz », « contextSwitch » and « clockFreq ». In the lower left-hand corner, we find an illustration of this « COntroller » class. One of its instances is called « myC ». The two non-functional constraints situated in the lower right-hand corner impact this myC instance. The first constraint is a « myC » property: its period of time required for changing the context is 8 μs (measured value) and its load limit is of 5%. The second constraint is of a contractual type: the working load limit of the process is higher than 90% if the rate of its clock is 60 MHz, and it is lower or equal to 90% if the rate of its clock is 20 MHz. Let us note that the two constraints are here expressed in VSL.

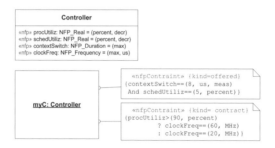

Figure 6.5. *Examples of real-time constraints*

6.3.2. *A components model for the real-time embedded system*

The objective of MARTE in terms of component-based approaches is not to redefine one of those new components models that are already so numerous. On the contrary, the main idea is to make a maximal reuse of what we have and to allow us to make the best use of an architectural MARTE model toward a maximum of possible platforms, such as AUTomotive Open System ARchitecture (AUTOSAR) or architecture analysis and design language (AADL).

The concept of components in UML is available in two forms defined in clauses 9 and 10, called « Components » and « Composites Structures ». Clause 9 extends the concept of UML class in order give it an internal structure, which is ultimately hierarchical, as well as ports. Such a class, often called « composite class », can be linked to its internal parts via « delegation connectors », linking the ports of the class to the ports of its internal parts. The internal parts of a composite class can also be assembled by connectors, then called « assembly connectors ». The connectors, whether delegation connectors or assembly connectors, specify solely the possible communication pathways between the components that constitute a system; the connectors do not specify the nature of the communication, as is the case, for example, in a synchronous or asynchronous communication.

Historically, UML is an object-oriented language. Because of this, the communication principle that has been preferred among the components of a system is the classic operations call, mainly synchronous or asynchronous, and that is included in a general model often called "client-server." UML also allows for a communication called "via signal". A signal can carry data and the sending of a signal is asynchronous and can be done in dissemination mode (*broadcast*) or multipoint emission (*multicast*). The set of concepts defined in clause 10 of the UML norm thus enables us to describe an architecture model completely.

Clause 9 is based, by extension, on the concepts of clause 10 and completes UML in terms of the architectural description, so that it favors the components reuse. The concepts defined in this clause allow us to see a component as a black box connected to a certain number of external artifacts. These can be, for example, the source code implementing the interfaces provided by the components, or even a documentation on its programming interface.

For MARTE, it is then naturally the composite structures of UML that have been chosen as its foundation for component-based modeling. In this context, the extensions proposed in the GCM subprofile have been purposely reduced to a minimum. GCM thus proposes concepts that enable communication by data exchange between the components, and a support facilitating the modeling of the interfaces provided and required in the framework of a client–server communication. Furthermore, a large contribution of the GCM profile is its precise description of the relations between the static and dynamic aspects of the internal semantic of the components. This clarification mainly consists of describing as precisely as possible what happens at the level of a component, or more precisely at the level of its ports, that is what happens when they receive a flow of information (reception of a data flow, of a signal or an operation call) in comparison to the internal behaviors of the composite class that has them.

In UML, the specification of the information that can travel via a class port, that is the specification of its provided and required interfaces, is defined by the type of port. Indeed, a UML port has natively two properties dedicated to the definition of its provided and required interfaces, but both properties are defined as derived, which means that their values are calculated. In this case, the set of interfaces provided is defined as the set interfaces carried out by the type of the port (plus the actual interface, if the type is itself an interface). The set of interfaces required is defined as the set of interfaces used by the type of port. If we consider the example in Figure 6.6, the « C » component has a port « p » that offers the interfaces « I1 » and « I2 » and requires the interface « I3 ». Indeed, class « P » that typifies the port « p » of « C » realizes the interface « I1 » and uses the interface « I2 ».

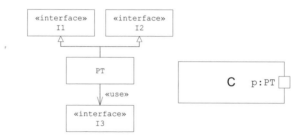

Figure 6.6. *Example of the specification of interfaces provided and required of a UML component*

Figure 6.7 illustrates the specific notation proposed by UML for graphically visualizing the interfaces required by and provided for a port.

Figure 6.7. *UML specific notation for the required and provided interfaces*

The main inconvenience of this method for specifying provided and required specification interfaces of a port is that it needs the modeling of an intermediary construction in order to define the modeling element that will be used to typify the port. As previously discussed, this element must realize the interfaces that we wish to provide throughout a port and use the interfaces required.

This aspect having been considered cumbersome by a significant number of users, MARTE proposes to briefly present the syntax throughout the stereotype « ClientServerPort » annotating the UML ports. The « provInterface [*] » and « reqInterface [*] » properties of the stereotype allow us to specify more explicitly the list of required and provided interfaces. In this case, the port must not be typified. Figure 6.8 shows how MARTE enables us to specify the same thing as shown in Figure 6.6: port « p » is now stereotyped as « clientServerPort » and its properties « provInterface » and « reqInterface » were informed in such a way as to specify that it proposes the interfaces « I1 » and « I2 », and that it requires the interface « I3 ».

Figure 6.8. *Example of a client–server port based on MARTE interfaces*

In addition, MARTE proposes another means for defining the ports via one single interface, specifying directly and simultaneously the services offered and the services provided by the port. In this case, the port is also stereotyped « ClientServerPort ». This stereotype has a particular property called *featureSpec* that allows us to reference a particular interface. This interface must be stereotyped « ClientServerSpecification ». An interface then has operations (the services) and/or receptions declaring the signals that can be perceived. These behavioral characteristics are then annotated with the stereotype « ClientServerFeature » whose property « kind » determines if it is offered, required, or both at the same time. In this case, the respective values of the property are *provided*, *required* or *proreq*. Figure 6.9 shows an example of using this situation: port « p » is specified by the interface « I1 » that defines that it offers the service « service1 » and requires the service « service2 ».

Besides the previously presented client–server communication model, MARTE facilitates the description of an architecture based on components that communicate with each other via data exchange. This type of model is possible in UML but uniquely at the level of the behavioral description of a system. To complete the UML components model, and similarly to both SysML and AUTOSAR, MARTE extends the concept of a UML port by introducing the concept of data port throughout the stereotype « FlowPort ». Just like for the client–server ports, the specification of the information traveling via a data port is done throughout its type. What is more, the data ports can be atomic or not. In the first case, their type is either a UML signal, whether a primitive type, or a type of structured data. In the second case, the port is typified by a UML interface, which is itself stereotyped by « FlowSpecification ».

Figure 6.9. *Example of a port defined by a « clientServerSpecification »*
MARTE interface

The properties of a data port, that is a stereotyped « FlowPort » UML port, are thus a Boolean number called « isAtomic », which we have just talked about, and the property called « direction » whose possible values are defined by the enumerated « FlowDirectionKind » (*in*, *out* or *inout*). If a value is fixed for this property, then the direction of all the data traveling via the port and that are specified by its type must be in accordance with the value of the property.

The model of MARTE components authorizes the two forms found in data communication, the active form and the passive form[3]. The first one is called an active form because the reception of the data triggers the execution of a behavior that

---

3 The active form is often called *event-triggered* or *push-based*, whereas the passive form is called *time-triggered* or *pull-based*.

is associated with the receiving component. In this case, the received data are not stored by the component but sent to the execution of the triggered behavior. This first form is mainly based on the UML event causality model. In the second case, the passive form, the reception of the data does not trigger any activity in the middle of the component. The data are stored in an *ad hoc* structure of the component and are ultimately read throughout the execution of a behavior of the component triggered by another means (e.g. a behavior whose execution is periodic). In the case of the passive form, the data stored are not consumed, which does not mean that they are deleted, but that they are only read. Consequently, such data can be read in several rounds, by the same execution or by different executions. This second form of semantics rests on a particular use of the properties and on UML connectors.

6.4. Conclusion

The aim of this chapter was not to undertake a systematic review of the concepts defined in the MARTE norm but to give you, the reader, a sufficient level of information for best understanding the rest of the chapters in this book, and the necessary information to better and efficiently read the specification itself, thus to go further. Indeed, the following three chapters will focus on demonstrating several uses and concepts of MARTE on the example of the pacemaker, presenting them in more concrete detail.

6.5. Bibliography

[CEA 05] CEA, The Carroll Research Programme, 2005. Available at www.carroll-research.org/uk/index.htm.

[ESP 09] ESPINOZA H., CANCILA D., SELIC B., GÉRARD S., "Challenges in combining SysML and MARTE for model-based design of embedded systems", *Proceedings of the 5th European Conference on Model Driven Architecture – Foundations and Applications*, Enschede, The Nethderlands, 2009.

[GER 03] GÉRARD S., TERRIER F., *UML for Real-time: Which Native Concepts to Use?*, Kluwer Academic Publishers, The Netherlands, 2003.

[OMG 05a] OMG, OMG UML profile for schedulability, performance, and time, Version 1.1, OMG Document formal/2005-01-02, 2005.

[OMG 05b] OMG, UML profile for MARTE: modeling and analysis of real-time embedded systems, RfP, OMG document realtime/05-02-06, 2005.

[OMG 09a] OMG, OMG policies and procedures, Version 2.9, OMG Document pp/09-12-01, 2009.

[OMG 09b] OMG, UML profile for MARTE: modeling and analysis of real-time embedded systems, Version 1.0, OMG document ptc/2009-11-02, 2009.

Chapter 7

Case Study Modeling Using MARTE

7.1. Introduction

In this chapter, we will introduce the modeling of the case study of a pacemaker, using the UML-MARTE notation. Having laid out the MARTE principles in Chapter 6, in this current chapter we will assume the reader has understood them.

7.1.1. *Hypotheses used in modeling*

Since this study is based on the general description of a pacemaker [BOS 07], on its basic functioning principles [BAR 10] as well as on Chapter 2, we will consider these documents as subsections of the requirements specification of our case study.

We have also assumed that the case study previously modeled in SYSML in Chapter 4 is the system analysis specification. We will detail it, in the current chapter, in a software analysis specification.

From the case study in question, the modeling only deals with the part called "real-time embedded software". Therefore, we will only present the part that concerns the pulse generator. Given that the monitoring and configuring station (called *Device Controller Monitor* in system specifications) will not be implanted in the body of the patient and thus has less "real-time" constraints, we have not deemed it necessary to describe it in this chapter.

Chapter written by Jérôme DELATOUR and Joël CHAMPEAU.

7.1.2. *The modeling methodology used*

MARTE is simply a standardized notation (i.e. a UML notation profile); it therefore allows the users to choose their own methodology and the set of practices they are accustomed to in order to define and carry out their own process.

The UML language is extremely rich due to the different diagrams it uses to describe the different viewpoints of a software system. Given that a model represents an abstraction of a real system, and has a very precise intention, these different viewpoints are crucial for obtaining a good abstraction of the system. Therefore, in what follows, we will introduce our modeling of the case study, explaining what we wish to cover in this chapter.

This set of intentions has to be well organized, thus constituting a modeling methodology. This methodology can be applied by using UML/MARTE and, in this case, every modeling intention will be associated with one or several diagrams. The couples intention and the UML diagram allow us to define a process of using UML language, which is indispensable to the actual use of such a language, as its sphere of applicability is very wide.

This is why we will present what we are looking to address in order to describe the different modeling stages. However, this modeling process does not intend to serve as a reference methodology for all real-time system developments. Moreover, for pedagogical reasons and given the context, we will only present a simplified version of this modeling process.

Throughout our modeling stages, we will take into account the need for model validation and code generation which will be performed in the following stages of the process we have described.

Within this methodology, we have started with the SysML modeling and certain system specifications that we have laid out in Chapter 4. Then, we have refined our model in order to explain the following intentions:

– of the system, viewed as a "black box system" (i.e. the content of the system is not detailed). This software analysis is based on several use cases that describe different scenarios applying to the system.

– Modeling the software architecture based on a so-called behavioral approach. The software design (called "preliminary design") which enables us to come up with an architecture in the form of a diagram of structural classes, allows us to grasp all the behavioral scenarios using a glass box type of analysis (i.e. the content of the system is detailed). This intention can itself be further broken down into two intentions:

- the first intention describes the proposed architecture that represents the breaking down of the system into collaborating entities;

- the second intention is a behavioral description of each of these entities that are internal to the system.

Formal analyses have been performed starting from these design models. We will lay out these analyses in Chapter 8. Next, a detailed design was carried out that has allowed us to generate a code for this case study. This part (detailed design and code generation) will be described in Chapter 9.

This modeling process has been iterative, demanding several iterations of these different stages (software specification, software design, formal analysis and detailed design). Some of the transitions between these activities are automated (toward the formal analysis and code generation). These automations rest on different transformational techniques and are actually assigned to different software engineering tools. Therefore, the reader should not be surprised to find UML/MARTE illustrations being founded on different modeling tools.

7.1.3. *Chapter layout*

The current chapter solely describes the phases of analysis and software design that make up a system. We will therefore start with the software analysis phase using a black box type of analysis, and then we will turn our attention to the architectural design based on a glass box type of analysis of the interactions between objects. Lastly, we will finish with a description of the internal behavior of various entities of the architecture.

7.2. Software analysis

The software specification groups together the black box analysis of the system starting from use cases and from the detailed interactions between the system and the actors that are external to the system. This stage allows us to characterize the interfaces of the system on the basis of the behaviors that are expected from the system.

7.2.1. *Use case and interface characterization*

There is ample literature on the methodologies associated with UML use cases [COC 01]. However, in this work we will aim to suppress things in order to focus on an approach that will be especially useful as an introduction to the stages of model behavior validation. Although the use cases focus on the functional requirements of the system, they represent a main framework for defining the system model. They regroup and abstract all the necessary behaviors and propose an organization that will then serve as a guide for carrying out the interaction scenarios, for identifying internal entities as well as for the validation phases that rest on the validation of software model requirements.

Starting from use cases that have been established at the level of SysML modeling, we will refine even further the use case *Routine Followup*. This use case is related to three external actors, that is the *Physician*, the *Technician* and the *Patient*. This use case is detailed by explaining the smaller use cases it incorporates, in order to characterize the larger interactions in the functioning of the *Pacemaker* system. In the use case, the actors *Physician* and *Technician* are identified as having precise objectives in relation to the system and are thus actors that trigger interactions toward the system. Therefore, they are defined as the main actors. The *Patient* is a secondary actor because it will have to interact with the *Pacemaker* system in order for the system to provide the service required by the main actors.

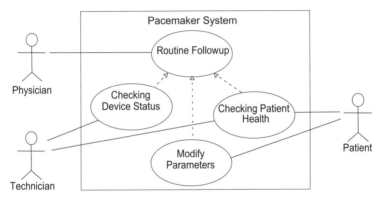

Figure 7.1. *Refinement of the use case Followup*

The objective of both the *Physician* and the *Technician* is to reach a state of well-being for the *Patient* using the *Pacemaker* system that will provide the data it picks up from the *Patient*. In order to achieve this goal, the *Physician* or the *Technician* triggers the interaction and the system provides the service after having picked up, stored and processed the data that come from the *Patient*.

The use cases include an internal textual description that defines the context and the scenarios relative to this use case. This internal description is based on different reasons, and we have chosen to limit the definition by incorporating the objective, a textual description, the actors involved in this use case, the event that triggered the case and at least one reference to a scenario that describes the nominal behavior of the case (see Figure 7.2).

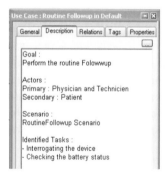

Figure 7.2. *Description of the use case Followup*

By looking to identify the sub-objectives comprised in *RoutineFollowup*, we will be able to associate these objectives to three use cases: *CheckingDeviceStatus*, *ModifyParameters* and *CheckingPatientHealth*. Figure 7.1 shows three use cases that illustrate the case of general use *RoutineFollowup*. In this diagram, the relations between the use cases and their actors help us pinpoint the actors that show up in the scenarios associated with use cases.

The refinement of this use case rests on the use of stereotyped dependencies *Extend* because these use cases are seen as optional in relation to the main use case *RoutineFollowup*. Once the pacemaker is installed, the *Physician* and the *Technician* indeed choose to trigger the interaction that corresponds to each of the use cases. The responsibility of the *Physician* covers all of the refined use cases that follow after the *RoutineFollowup* case. By contrast, the responsibility of the *Technician* is limited to two use cases: *Checking Device Status* and *Checking Patient Health*, which ultimately translate into relations between the actors and the respective use cases.

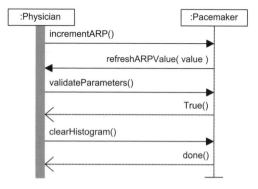

Figure 7.3. *Scenario related to the use case Followup*

A nominal scenario describes the interactions for each use case, especially for the case that describes the incremental changes in ARP value, which are in turn described by the sequence diagram in Figure 7.3. This modification can happen within the parameters of the pacemaker. The sequence diagrams per use case identify the messages that come in and go out of the system. Let us note that this scenario will serve as a basis for the breakdown of the system in objects that will later enable us to identify the system classes. Failure scenarios can also be added in order to complete the description of the interactions by identifying the steps that are to be taken if the system fails.

7.2.2. The sphere of application

Just like SysML, MARTE enables us to define different types of data. As opposed to SysML, MARTE provides, by default, a large library of common types within the domain of embedded software systems, whether it is for defining functional requirements or extra functional requirements. Thus, there is no need for us to define the type of time (as we have already mentioned in Chapter 6, MARTE disposes of a rich semantics for describing time constraints as well as the kind of time we are dealing with).

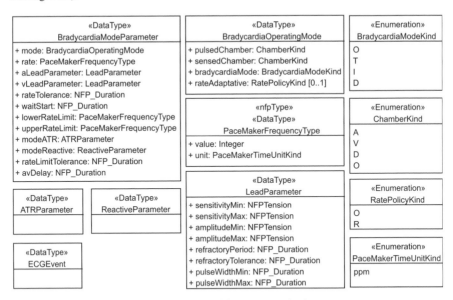

Figure 7.4. *Definition of domain-specific data types*

On the other hand, there are no descriptions of domain units of cardiac regulation in MARTE (such as the notion of pulsation per minute or operational modes).

Figure 7.4 presents the definition of domain-specific data types. We have already given the explanations regarding the definition of new units in Chapter 6.

For the medical definition of pulsations per minute, we have created a type of data *PaceMakerFrequencyType* and a type of unit (*PaceMakerTimeUnitKind*) to facilitate the expression of several requirements with the aid of this variable. Figure 7.4 also showcases the definition of the parameters of operational modes, namely *rateTolerance, waitStart, avDelay*, etc. These are defined by using the predetermined *NFP_Duration* type to express duration, while others (*lowerRateLimit, rate* and *upperRateLimit*) use the recently created *PaceMakerFrequencyType*.

7.3. Preliminary software design – the architectural component

System software design consists of breaking down the system into separate components (software and/or hardware components) that, when collaborating, take into account all of the analysis requirements.

At this stage of the modeling process, it is often assumed that the future system has infinite resources (in terms of memory, computing power, etc.). Here, we want to ensure that we achieve an architecture that is coherent, logical and complete. As for constraints such as parallelism and real-time, what we need to do is mainly identify the logical parallelism (i.e. the requirements demanded by the analysis).

The implementation details (such as the management of persistent data, the booting and shutting down of the system, the management of communications between the hardware nodes and the management of stimuli and responses) are generally either ignored or simplified. However, if these constraints are imposed by the requirements, they must be represented at this stage of the preliminary design (although there is no need to process them in detail). We will only address these aspects at a later stage of the detailed design.

For the sake of simplification, here are the main stages that the system design has to follow:

– Defining the candidate architecture, which allows us to reach the first stages of the system breakdown into collaborative components.

– Modeling behaviors, which allows us to define the behavior that is required of each component.

This is an iterative and recursive process. For purely pedagogical reasons, we will introduce the result of this process, without going into further details as to the different iterations that have been necessary along the way.

7.3.1. *The candidate architecture*

We will summarize the breakdown of the proposed system through SYSML modeling, keeping only the software aspects that are necessary for our development.

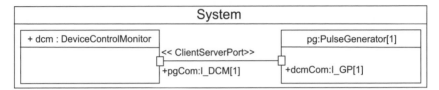

Figure 7.5. *Diagram of the general architecture of Pacemaker*

This view is represented in Figure 7.5 by a composite structure diagram.

This diagram as well as the semantics of its ports have been presented in Chapter 10. However, it is very similar to the internal block diagram of SYSML (laid out in Chapter 6) and is based on similar concepts and representations.

As previously indicated, we will only detail the PulseGenerator. The SYSML modeling has otherwise already identified a proposition that breaks down this component in subcomponents (PGControler, Battery and Lead) without specifying the nature of the said subcomponents (namely whether it is a software component, a mechanical component or a hardware component).

7.3.2. *Identifying the components*

In order to establish the candidate architecture and identify its components, we have mainly used the so-called *Class Responsibility Card* (CRC) technique [BEC 89]. Starting from interaction diagrams associated with the use cases written in the analysis phase, we have identified the internal components of the system and we are detailing their interactions in order to ensure that every message directed at the system is correctly processed by the elements that make up the system.

This stage is very important as it identifies the interactions between the different components. The identified interactions enable us to describe the interface (i.e. a set of operations) of each component. Therefore, in the example depicted in Figure 7.6, we have described a scenario of installing an operating mode. Let us notice the messages *switchOn* and *startMode* that are then carried over to the interface definition of the following paragraphs.

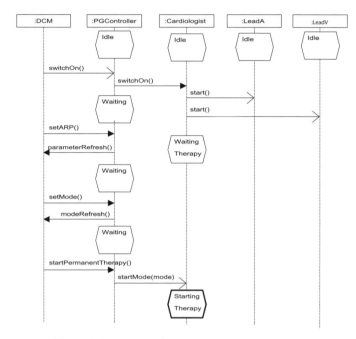

Figure 7.6. *Sequence diagram between system objects*

In addition these scenarios will be then used for modeling the internal behavior of each component. In fact, every incoming message received by an object needs to act as a trigger for the transition of a state diagram and every outgoing message that comes out of the object must be emitted in the same state diagram.

During this stage, where we identify the interactions, we may also express the constraints that are seen as properties the system must observe. In this framework, the use of MARTE is completely justified and necessary for obtaining non-ambiguous constraints that we can then interpret throughout certain stages of formal validation. These will be tackled in Chapter 8.

In Figure 7.7, we specify a timing constraint between two pulses sensed by the ventricle lead *LeadV*. This constraint is expressed on the basis of two instants defined by @*t1* and @*t2*. These instants are identified as being the instants when the messages are consumed *sensedPulse( )* by the *Cardiologist* type. This definition is made by using the stereotype *"TimedInstantObservation"* and by specifying the type of observation *obsKind* to *consume*. This definition leaves no ambiguity between the receipt of the message and its consumption.

With regard to the actual constraint, this is stereotyped *"TimedConstraint"*, also stating that the type is a constraint imposed on a period of time with *duration*. We hereby restate the difference between the instances that were previously defined as

@*t1* and @*t2*. These definitions of instants and constraints are made possible by the use of an idealClock. This idealClock, defined in the MARTE profile, refers to an ideal global chronometric clock.

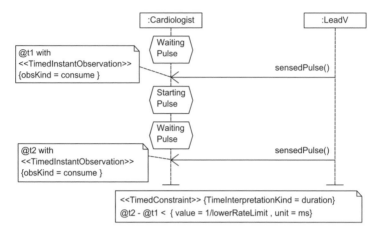

Figure 7.7. *Sequence diagram with MARTE time constraint*

Different scenarios thus become communicable in order to specify all the properties that the system needs to observe. Subsequently, these properties can be formally validated on the basis of modeling the behavior of the system components.

7.3.3. *Presentation of the candidate architecture*

Applying the CRC technique along with good design principles allows us to define a candidate architecture by identifying the subcomponents of the pulse generator. This candidate architecture is described by Figure 7.8.

The main responsibilities of the components are as follows:

– The controller (an instance of PGController) is in charge of communicating with the Device Control Monitor (DCM), along with applying the requests of the DCM and sending the responses over to the DCM. Through a means that is not specified in the requirements specification, it detects the proximity of a magnet.

– The cardio (an instance of Cardiologist) applies a policy of monitoring the bradycardia (see description of operating modes in Chapter 2), which is requested by the controller. Let us note that this component is a modeling entity and does not indicate the cardiologist that will use the system. This is a common practice in object-oriented modeling to reproduce the reality. Thus, in the rest of this chapter, when we speak of a cardiologist, of leads or batteries, we will in fact be referring to modeling entities.

– The leads, which can be identified according to their position on the atrium (atrium, *leadA*) or the ventricle (*leadV*), detect the heartbeats and can generate a simulation of the cardiac rhythm at the request of the cardio.

– The battery (an instance of Battery) manages the autonomy of the controller and can make the call for operating in safe mode when its charge decreases below a certain threshold.

– The logger (an instance of ECGLogger) preserves all the cardiac events that resend the leads pulsations and stimulations in order to generate an electrocardiogram (ECG).

– The accelerometer (an instance of Accelerometer) warns the cardiologist when it detects a more significant physical activity in the patient.

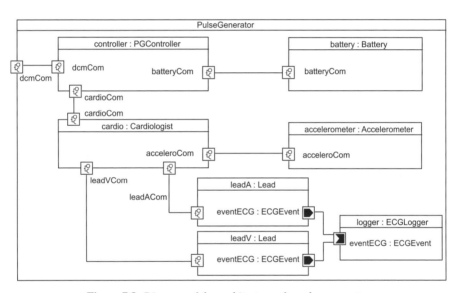

Figure 7.8. *Diagram of the architecture of a pulse generator*

For reasons of clarity, we made the choice to represent only the ports in Figure 7.8 as well as the main connections between components (e.g. the connections between the logger and the controller have not been represented). The connections between the leads and the logger are in fact data flows (*flowPort*). In this stage of the design, we will not yet specify if the flow is passive or active. We will only model the fact that we keep track of data sent by the leads. In fact, the logger must keep all of the signals whether they are sent (cardiac simulation) or merely detected (heartbeats) by the leads before sending out ECGs on demand on the monitoring station.

The other connections, such as those between the cardio and the controller, are connected to client–server ports. They are made up of interfaces based on offer and demand. These aspects will be treated in more detail in the next paragraph.

At this stage of preliminary design, we do not yet know if the components are software components or hardware components or a mixture of the two. For example, as far as the leads are concerned, it is highly probable that they will be broken down into both a software and a hardware subset; there will be a software part, which will run on an embedded card, and an electronic part, placed on the myocardium of the patient, which will communicate with the embedded card via electric signals.

7.3.4. *A presentation of the detailed interfaces*

We have chosen to represent only a part of the interfaces that are offered and demanded by different components, namely those of the interfaces offered and demanded by the component cardiologist *Cardio*; they are represented in Figure 7.9.

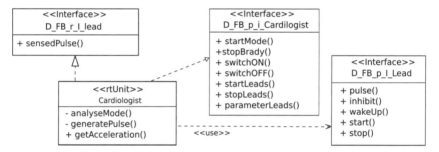

Figure 7.9. *Partial view of the interfaces offered and demanded by the cardiologist*

These interfaces allow us to identify the possible communications between components.

We have a first interface *D_FeatureBasedCS_r_I_Lead* offered by the Cardiologist (the *Cardio* component), which allows us to signal the pulse detection by the Leads (operation *SensedPulse*).

The cardiologist uses the required interface *D_FeatureBasedCS_p_I_Lead* in order to communicate with the leads. It can also ask for the generation of a cardiac stimulation (operation *pulse*), for the deactivation of the lead (operation *inhibit*) or for its reactivation (*wakeUp* method). The deactivation of the lead causes it to stop warning the *Cardiologist* component that a pulse has been generated. In fact, following the possible regulating operating modes, the decision of simulating the heart of the patient can be made by listening to one or the other of the leads.

Deactivating a lead thus enables the cardiologist to be selectively alert of a pulse only by the good lead. The operations *start* and *stop* model the turning on or the switching off of a lead (the activation or the deactivation of cardiac signals toward the logger).

The other interface offered by the cardiologist *D_FeatureBasedCS_ p_I_Cardiologist* models the possible communications with the controller. We are mainly concerned with installation commands and with starting up a monitoring and heart assistance policy. These operations will be detailed in the section dedicated to behavior modeling.

7.4. Software preliminary design – behavioral component

The objective of this modeling step is to define the internal behavior of each component. This behavior can be modeled by UML state diagrams or UML activity diagrams.

When a component is said to be active it represents a control flow that is independent of the application. Its behavior is often modeled by a UML state diagram. In MARTE, an active component is identified by the "RtUnit" stereotype. A *Real-time Unit* (RtUnit) is an entity that has its own control thread and may have more operating modes.

The behavior of the component is modeled via interaction diagrams (such as the sequence diagram in Figure 7.7); these diagrams were created during the previous stage of defining the candidate architecture. Moreover, we need to extend this behavioral description in order to ensure that the behavior of each component covers all the possible scenarios of using the system.

In this case study, most of the components are active. In the following, we will describe the behavior of the main components.

7.4.1. *The controller*

The controller is the main controlling entity because it is in charge of the different operating modes of the pulse monitor. There are five main functioning modes (represented in Figure 7.10):

– The *TemporaryBradyPacing* mode, which allows us to test and set the parameters of the monitor. This is a functioning mode that is mainly used in the implantation and calibration phases of the pacemaker.

– The *PermanentBrading* mode, which corresponds to the nominal functioning mode of the monitor (in outpatient cases and patient follow-up cases). An operational mode, tailored to the patient's needs, is applied.

– The *PaceNow* mode, which is used in case of emergency. A particular operating mode (called "VVI") is applied;

– The "*PowerOnReset*" mode, which is set off when the battery runs down and its capacity decreases under a certain threshold. Even then, the operating mode needs to be applied "VVI" and certain functions of the monitor will be deactivated.

– The "*Magnet* mode", which is a testing mode that checks the level of the battery of the monitor. Depending on the battery level and the operational mode then engaged in nominal functioning, a specific operational mode is set off by the regulator. This mode is set off by the approaching of a magnet near the area where the monitor is implanted in the patient.

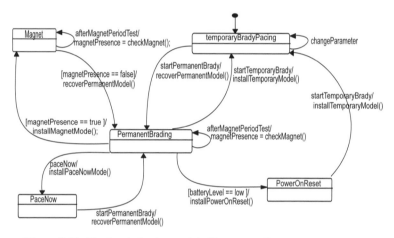

Figure 7.10. *Monitor behavior modeled by a state diagram*

For reasons of clarity, we have only represented the most important transitions in the state diagram in Figure 7.10.

In the *TemporaryBradyPacing*" mode, we have only described two transitions, one that enables us to set the parameters for different operating modes attached to different functioning modes. The second one shows the transition from the functioning mode *TemporaryBradyPacing* to the *PermanentBrading* mode. This is done by receiving the message *startPermanentBrady* and it triggers the request for performing a private method (*revoverPermanentMode*) of the monitor. At this phase of preliminary design, we assume that the sending of the message *startPermanentBrady* is performed by the device control monitor (*DeviceControlMonitor*) without being any more precise to the nature of this message (synchronized or not). This simplifying hypothesis will be abandoned once we reach the stage of detailed design. We will then have to model all the components necessary for the setting up of a wireless communication between the monitoring station and the monitor.

In the *PermanentBrading* mode, we test periodically the detection of the presence of a magnet (*MagnetPeriodTest*). If a magnet is sensed, the monitor will switch to *Magnet* mode.

7.4.2. *The cardiologist*

The cardiologist applies an operational mode that is demanded by the monitor. Therefore, it does not know anything about the different functioning modes of the monitor. Its behavior is represented in Figure 7.11.

In its initial state (*idle*), the cardiologist is inactive. Once started (by invoking its method *swithOn*), it will activate the leads (*startLeads* method) so that they start observing the pulse and notify the monitor. The cardiologist is then in *WaitingTherapy* state and waits for the monitor to ask it to follow a given operational mode. The application of an operational mode (by using the *startMode* method) is done in two stages. First, we need to set the parameters (with the method *analyzeMode*) and to follow the required mode (activation or inhibition of the leads, defining the parameters of the operating mode, etc.); then, depending on the required operating mode, it will enter in the corresponding composite state. Each of these states is described in the following section.

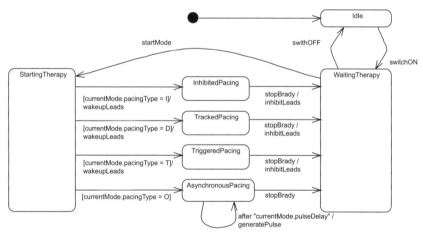

Figure 7.11. *Behavior of the cardiologist modeled by a state diagram*

7.4.3. *The operating modes of the cardiologist*

The operating modes can be broken down into four large strategies: asynchronous, triggered, inhibited or tracked. These different operating modes are explained in more detail in Chapter 2.

Operating mode with asynchronous strategy. The simplest of the regulating strategies, "asynchronous" (*AsynchronousPacing*), requires that the monitor sends periodically (*pulseDelay* period) a cardiac stimulation via the intermediaries of either one or both leads (following the corresponding operating mode "XXOX"). This is the internal private method *generatePulse*, which identifies what leads must send the cardiac stimulation.

Operating mode with triggered strategy. The so-called (*TriggeredPacing*) strategy consists of generating a cardiac stimulation as soon as a pulse has been sensed by one of the leads. The detection of the current pulse can only be made after a short delay (*RefractoryPeriod*, noted as RP in the diagram), in order to avoid the currently sensed pulse getting distorted by noises from the previous stimulation, as well as from the next pulse.

There are different operating sub-modes that can identify which lead senses the pulse and which lead detects the electric stimulation. However, due to the parameters of the mode that had been performed before entering this composed state, the description given in Figure 7.12 remains identical, regardless of the operating sub-mode that has been deployed. The activation of the leads (using the lead that must warn the cardiologist of having picked up a pulse) is indeed made by the *wakeupLeads* method. Similarly to the so-called "asynchronous" strategy, it is the *generatePulse* method that determines what lead sends the cardiac simulation. This observation stands for all operating modes that we will describe hereon in.

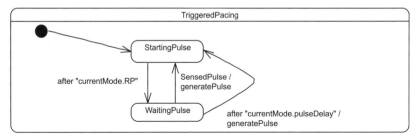

Figure 7.12. *State diagram for the composed state TriggeredPacing of the cardiologist*

Operating mode with inhibited strategy. The strategy called "inhibited" (*InhibitedPacing*) is close to the "triggered" strategy. As shown in Figure 7.13, if the heart starts beating on its own, we will not generate any cardiac simulation. Otherwise, we will generate a cardiac simulation using a lead after a certain period of time (*pulseDelay*).

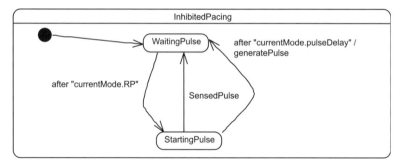

Figure 7.13. *State diagram of the composed state InhibitedPacing*
of the cardiologist

Operating mode with tracked strategy. The strategy called "tracked"
(*TrackedPacing*) is the most complex strategy so far (see Figure 7.14). A cardiac
stimulation in the ventricle after a fixed delay *AVDelay* must follow the pulse sensed
in the lead (*sensedPulsed ON LeadA*), unless, in the meantime, a pulse has been
sensed *sensedPulsed* in the ventricle. However, if no pulse has been detected in the
lead after a fixed period of time (*pulseDelay*), we must then generate a cardiac
stimulation in the ventricle.

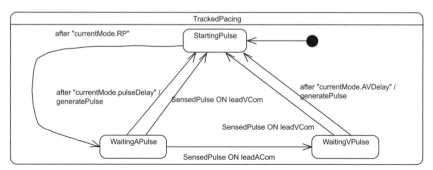

Figure 7.14. *State diagram of the composed state TrackedPacing*
of the cardiologist

7.5. Conclusion

This chapter provides UML/MARTE models of the preliminary design phase of
the pacemaker. We have tried to show how to use the UML/MARTE notation for this
stage, starting from different system use cases, and ending with obtaining the
candidate architecture of the system, including the behavioral description of the
components of the system.

The use of MARTE has enabled us to define the cooperation and collaboration of the system's components. Precise timing constraints and a behavioral description of the components have been modeled. The level of description is far more accurate in order to allow subsequent formal verification. These preliminary design models are then detailed in order to prepare the phase of code generation for a specific platform. They will also be used as a grounding for the formal verification described in Chapter 8.

7.6. Bibliography

[BAR 10] BAROLD S., STROOBANDT R., SINNAEVE A., *Cardiac Pacemakers and Resynchronization Step by Step: An Illustrated Guide*, John Wiley & Sons, Hoboken, NJ, 2010.

[BEC 89] BECK K., CUNNINGHAM W., "A laboratory for teaching object oriented thinking", *ACM SIGPLAN Notices*, vol. 24, pp. 1–6, September 1989.

[BOS 07] BOSTON SCIENTIFIC, PACEMAKER System Specification, January 2007.

[COC 01] COCKBURN A., *Writing Effective Use Cases*, Addison-Wesley Professional, 2001.

Chapter 8

Model-Based Analysis

8.1. Introduction

Nowadays embedded computer systems have become more and more widespread among the automatic devices in our daily lives. We find them in various transport systems: automobiles (global positioning system (GPS), braking systems, engine control, etc.), avionics (navigation system and flight control systems, pressure control and air conditioning, etc.), in medical equipment (monitoring patients, controlling the functioning of devices such as a pacemaker) and in construction building (fire safety, elevator control, power management systems, etc.).

This type of software ensures automatic functions, crucial functions even, because sometimes, human lives as well as highly important economic stakes are at risk. It is thus very important to ensure the good functioning of embedded computer systems, that is to make sure that there are no errors or bugs when compared with their functional specifications. As a result, finding the bugs and ultimately fixing the system in accordance with the system specs have always been two very significant phases in the software engineering cycle, ever since software systems have been first industrialized.

In the past, as long as embedded computer systems remained relatively simple (that is as long as their inner working, as well as the environment they were meant to control, remained simple), the task of finding the bug could be done manually. The main body of systems that had to be validated was, in this case, often designed and implemented as a sequential and deterministic software that would run in a loop

Chapter written by Frédéric BONIOL, Philippe DHAUSSY, Luka LE ROUX and Jean-Charles ROGER.

when interacting with a sampled environment. The validation phase was carried out through a series of tests or simulations. However, the fact that the capacity of these systems has improved gradually and also that they have become more and more widespread has lead, on the contrary, to an irreversible increase in their complexity. These complexities can be divided into two categories.

– First, there is an external complexity connected to the environment that the embedded system must control. This environment is indeed more and more often made up of several distinct yet interdependent physical entities, entities that must be controlled coherently and simultaneously. This is, for instance, the case for flight control systems that need to manage more than 20 control surfaces at the same time, being thus connected to more than 30 sensors.

– Second, there is an internal complexity that is connected to the software architecture of the embedded system.

This architecture is also more and more often composed of several software processes that interact with each other at the same time.

In both cases (internal and external), the simultaneity that characterizes the entities or the processes leads to an explosion in the number of possible behaviors; this, in turn, leads to an explosion of test scenarios that must be looked into before validating the system. To test this explosion, it suffices to consider a system made up of 10 actors that interact simultaneously, each of them carrying out a simple sequence of nine actions. Supposing an extreme scenario where the actors are completely asynchronous (i.e. there is no synchronization between the actors), then the number of possible combinations between the 90 actions, and therefore the number of possible cases of running the system in such a way as to validate it, becomes of the order of 3×10^{82}, which is almost equal to the number of atoms in the universe.

Obviously, totally asynchronous systems or environments do not in fact exist, since the actors and their actions are often at least partially in synch. It remains, however, true that real systems are often made up of different processes (of the order of tens) and they execute a large number of actions (often much more than nine), and also that the environment they have to control can also be composed of several entities (usually, of several tens of entities). The number of potential behaviors to be validated thus becomes enormous. This phenomenon is known as a "combinatorial explosion". Its immediate consequence is that manual testing methods and simulations are no longer adequate.

As a result of this observation, researchers have explored several techniques; among these, the family of formal methods that have, over the years, added to the already existing body of competent and accurate solutions for helping designers think up non-faulty systems. As we have seen in Chapter 1, in this field, *model-checking* techniques [QUE 82, CLA 86] have been strongly promoted because of their ability to automatically carry out property tests on the software models. To this end, several tools (*model-checkers*) have been developed [FER 96, HOL 97, LAR 97, BER 04].

Broadly, these tools try to model in a compact and abstract fashion the set of possible behaviors of the environment as well as of the system to be validated; exhaustively carrying out each such scenario and finally, as a result, deciding if the set of possible executions is faultless. The quickness with which this entire process is carried out depends on the compaction degree of the set of behaviors. A great deal of research has been carried out in this respect [VAL 91, MCM 92, GOD 96, EME 97, ALU 97, PEL 98, BOS 05, PAR 06]. However, given the enormous size of the sets considered (often several orders of size higher than the number of atoms in the universe), the real progress of monitoring tools does not yet allow us to process the real, industrial-sized systems.

In this chapter, we will lay out a complementary path, one that relies on *model-checking*, allowing us to push forward the limits of combinatorial explosion. This path seeks to limit the two sources of complexity: external complexity (i.e. that of the environment) and internal complexity (i.e. that of the system). More precisely, this approach seeks to reduce the space of possible behaviors by considering an explicit model of the system's environment. In other words, we seek to "close" the system with the set of use cases relative to the environment it is supposed to respond to, and of this environment only. The reduction is based on a formal description of these use cases that the systems interact with, hereby called *contexts*.

The objective is thus to guide the model-checker in focusing its efforts not on exploring the entire space of behaviors – which may be enormous – but on validly restricting it in order to check for specific properties. This property check therefore becomes possible in the state space thus reduced. This method is founded on the designers' knowledge and on their expertise, which allows them to explicitly specify the environment of the system.

As with every formal method, the approach we present here rests on a set of formal languages, a set of links between one language and the other as well as on exploring tools and monitoring tools. In addition, we will develop the principles of this approach by using two particular languages: Fiacre [FAR 08], for describing the model system that needs to be validated, and *context description language* (CDL), for describing the use cases of the environment considered for validation.

We will then use three tools coupled with these two formalisms. The first one is the TINA SELT model-checker[1] [BER 04]. This is a model-checker that is particularly well adapted to the verification of the requirements that express invariants and that are formalized in temporal logic.

---

1 Developed at LAAS, www.laas.fr/tina.

The second one is an OBP Explorer[2], which is an explorer of a Fiacre model coupled with an accessibility analyzer. OBP Explorer is more adapted to the property checks that can be expressed via observers. Both of these tools are connected upstream to a third tool, OBP[3]; OBP applies the method of behavior space reduction presented in this chapter.

This method and these tools are illustrated in the case study of the Unified Modeling Language (UML) Modeling and Analysis of Real-Time and Embedded Systems (MARTE)-described pacemaker, which was presented in Chapter 7. The case study is converted in Fiacre (the input language of the two explorers/*model-checkers*), and then the performances of the two tools are studied over two types of requirements: an invariant (expressed in temporal logic) and an observer (expressed in CDL language).

We can then show that the TINA SELT and the OBP Explorer, if used without the context reduction method, cannot overcome the phenomenon of combinatorial explosion that is inherent to the case study of the pacemaker.

We will then show that taking into account use cases of the pacemaker's environment, formally modeled in CDL, allows us to considerably reduce this combinatorial explosion, using TINA SELT as our OBP Explorer. More precisely, we will show that breaking down the environment into separate use cases, completely covering the set of environment behaviors (i.e. partitioning it, forming a partition in the mathematical sense using OBP), allows us to push the limits of the combinatorial explosion further. We will then show that partitioning of the environment can start automatically each time the barrier of the combinatorial explosion is reached, thus allowing the property check to take place on large-scale models.

The chapter is organized as follows. In section 8.2, we will describe that part of the UML-MARTE model that will be considered for the transformation in Fiacre language, whose syntax will be presented briefly. We will choose two requirements that we wish to check on the model.

Section 8.3 describes the principle of requirement checking by considering use cases that correspond to the programming models of the pacemaker. Test results are shown for both tools: TINA SELT and OBP Explorer. The technique of reduction via context exploitation, using OBP, is described in section 8.4 as well as its implementation in CDL language. We present the verification method using both tools, TINA and OBP Explorer. This chapter ends with an assessment and a conclusion detailed in sections 8.5 and 8.6.

---

2 Developed at the ENSTA-Bretagne and available at www.obpcdl.org
3 From a practical point of view, OBP integrates the OBP Explorer tool.

8.2. Model and requirements to be verified

In this section, we will present the part of the pacemaker model translated in Fiacre starting from a UML-MARTE model. We will also broadly present the Fiacre language and the translation principles from UML to Fiacre. We will choose two requirements from the specifications of the pacemaker for presentation that will be affected by the check.

8.2.1. *The UML-MARTE model that needs to be translated in Fiacre*

The UML model, introduced in Chapter 7 is interpreted as a set of processes (see Figure 8.1) whose behaviors are described with the help of state machines. These processes communicate via *first-in first-out* (FIFO) links or via shared variables (registers).

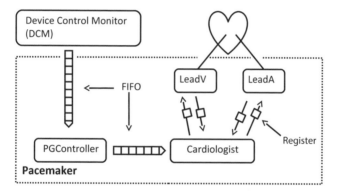

Figure 8.1. *Architectural model of the pacemaker*

The environment is composed of a *Device Control Monitor* (DCM) and the heart. The pacemaker is composed of the PGController that receives the instructions from DCM (modes and values of the parameters) and the Cardiologist that receives signals from the sensors LeadA and LeadV. These are stimulated by the heart. The Cardiologist takes up the responsibility of acting upon LeadA and LeadV according to the modes and parameters received from DCM via the PGController. The DCM, manipulated by the physician, sends parametering messages to PGController. The behavior of the PGController, LeadA and LeadV are modeled by the automata in Figure 8.2 and that of the Cardiologist process corresponds to the automaton described in Chapter 7 and in Figure 7.11.

The process PGController receives the messages from DCM, processes them before sending them out to the Cardiologist. It ensures the coherence of the messages (making sure there are no negative parameters or incoherent values) and sends the parameters over to the Cardiologist. To apply them to the

`Cardiologist`, it first requires that the currently running mode stops, sends out the new parameters and finally starts the new mode.

To apply the new parameters, the `Cardiologist` process must go back to the idle mode called $Waiting$. Once the mode is configured, the `Cardiologist` process oversees the pulse ($Pulse$) of the heart and applies a stimulation policy. The `Cardiologist` is capable of counting the time that separates the two heartbeats (a cycle) and to react in accordance with the functioning parameters.

The `Leads` transform the heartbeats in messages that are comprehensible for the `Cardiologist`. They are also in charge of applying the stimulation of the heart (signal $Pace$) at the command of the `Cardiologist`.

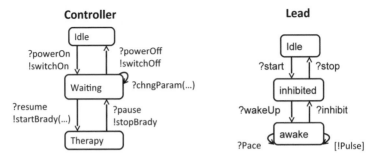

Figure 8.2. *Automata of the PGController and Lead*

In this model, we seek to model the behaviors by considering realistic scenarios, taking into account the behavior of the DCM and of the heart, as described in section 8.3.1.

8.2.2. *Fiacre language*

Fiacre language (Intermediate Format for the Architectures of Distributed Embedded Components) [FAR 08] has been developed within the Toolkit in Open Source for Critical Applications & Systems Development (TOPCASED) project as a key language linking high-level formalisms such as UML, Architecture Analysis and Design Language (AADL) and Specification and Description Language (SDL) with formal analysis tools. Using an intermediary formal language has the advantage of reducing the semantic gap between the high-level formalisms and the descriptions internally manipulated by verification tools such as Petri nets, process algebras or timed automata. Fiacre can be considered as a language disposing of a formal semantic that serves as the input language for different checking tools.

Fiacre is a formal specification language that describes the behavioral and timed aspects of real-time systems. It integrates the notions of process and components as follows:

– the automata (or processes) are described by a set of states, a list of transitions between these states with classical constraints (variable allocations, *if-elsif-else*, *while*, sequence compositions), non-deterministic constructions and communications done via ports and via shared variables;

– the components describe the compositions of the processes. A system is built as a parallel composition made up of components and/or processes that can communicate between them via ports or shared variables. The priorities and the time constraints are associated with communication operations.

The Fiacre processes can be synchronized with or without value passage via the ports. They can also exchange data between them via asynchronous communication queues using shared variables.

This is the communication mode that we use in the Fiacre model. The expression of time in Fiacre language is based on the semantic of timed transition systems (TTS) [HEN 91]. Every transition is associated with time constraints (a minimum time and a maximum time). These constraints ensure that the transitions can be done within the time intervals defined (not earlier nor later).

8.2.3. *The translation principles of the UML model in Fiacre*

These are the translation principles of the UML model in Fiacre language.

The active objects of the UML model are translated by instances of Fiacre process. They communicate with each other via shared variables. They enable the modeling of shared registers and of message threads by an asynchronous communication in FIFO mode.

According to the indications in Chapter 7, the objects PGController, Cardiologist, LeadV and LeadA are implanted by the Fiacre process with the same name. On the other hand, the objects accelerometer battery and logger do not bear any influence on the behavior of previous objects. Therefore, they have not been considered in the Fiacre modeling for the analysis of the model.

The Fiacre model has a *Pacemaker* component (Listing 8.1) that composes the instances of the processes running at the same time (operator ||): PGController, Cardiologist and two instances of the Lead process. We will add to the component an instance of the DCM process in order to simulate its behavior. These instances communicate via shared variables of the t_fifo type for the message threads or of t_registre type for the registers. Thus, the instances of DCM and PGController processes share the variable *DCMToController* declared as a message thread. The PGController and Cardiologist instances share the message thread *ControllerToCardio*. The Cardiologist instance shares with the first (respectively, with the second) Lead instance the shared variables *CardioToLeadA* and *LeadAToCardio*: (the variables *CardioToLeadV* and *LeadVToCardio*, respectively).

```
component Pacemaker is
var
DCMToController, ControllerToCardio : t_fifo,
CardioToLeadA, CardioToLeadV,
LeadAToCardio, LeadVToCardio : t_registre,
par
    DCM         (& DCMToController)
 || PGController (& DCMToController, & ControllerToCardio)
 || Cardiologist (& ControllerToCardio, & LeadAToCardio,
                  & CardioToLeadA, & LeadVToCardio,
                  & CardioToLeadV)
 || Lead         (& CardioToLeadA, & LeadAToCardio)    // instance leadA
 || Lead         (& CardioToLeadV, & LeadVToCardio)    // instance leadV
end
```

Listing 8.1. *Fiacre declaration of the Pacemaker component*

The behavior of each process is modeled by an automaton. The states and transitions of UML *state-charts* are translated by Fiacre states and transitions. Listing 8.2 partially showcases the programming of the process (`PGController`).

```
process PGController (& inputFromDCM : t_fifo_DCMToController,
                      & outputToCardio : t_fifo_ControllerToCardio) is
states
  Idle, Waiting, Therapy
var
parameters : t_parameters,    // An array of parameters (DCM level)
modification : t_modification, // A parameter index and its new value
  mode : t_Mode           // An array of  parameters (Cardio level)
init
      to Idle

from Idle
if not (empty inputFromDCM) then
  case first inputFromDCM of
    PowerOn ->
     inputFromDCM := dequeue inputFromDCM;
     outputToCardio := enqueue (outputToCardio, SwitchOn);
     to Waiting
      | any -> null // Should not happen
  end case
end if;
  loop

from Waiting
    ...

from Therapy
    ...
```

Listing 8.2. *Fiacre declaration of the PGController process*

For the asynchronous communications, the operations *first*, *dequeue* and *enqueue*, respectively, allow us to read a message at the beginning of a queue, to delete a message at the beginning of the queue and to write a message at the end of the queue. Every process assigns itself a unique entry queue. Writing a message in this queue (*enqueue*) corresponds to sending the message.

For this translation, the UML model is interpreted using the following semantic: the ordering of simultaneous processes (processes that take place at the same time) rests on the atomic and non-deterministic interleaving of the transitions of each processes ("atomic" meaning that a transition cannot be paused in the middle of its execution). In other words, for all processes that have fireable transitions, a transition may be chosen in a non-deterministic fashion and executed atomically. Then, another transition is chosen from the set of fireable transitions. This ordering policy is in accordance with the semantic hypothesis, followed in Chapter 7 throughout the UML modeling of the system of the pacemaker, and corresponds to the semantic of Fiacre programs.

As we have previously seen, the representation of time in Fiacre is symbolic, based on intervals. In section 8.2.4, we wish to check the requirements that refer to periods of time between two signals or during a cycle.

Given the requirements that we need to check, we must be able to manipulate the values corresponding to the time periods between the events or the actions carried out by a process. We choose to manipulate variables or meters that can measure the time by breaking it down into discrete units. The time unit chosen can be parametered.

8.2.4. *Requirements*

This section focuses on two requirements which have been introduced in Chapter 5. We wish to verify them on the model Fiacre. These requirements regard the parameters specified with the help of DCM.

The first requirement (requirement 8.1) regards the tracking mode (VDD). In the behavior of the pacemaker, an auricular signal is followed by a ventricular signal. The parameter *Atrial Ventricular Delay* (AVDelay) defines the maximum time interval that may pass between these two events. In the absence of the response expected from the ventricle, the pacemaker must stimulate it. Having received an auricular signal in tracking mode (VDD), the pacemaker need not "wait" for more AVDelay time units. In this model, we chose milliseconds as time units. The requirement is expressed thus as follows:

REQUIREMENT 8.1.– In VDD mode, the time between an auricular signal and a response or a stimulus coming from the ventricle cannot be higher than the value of the parameter AVDelay.

The second requirement (requirement 8.2) describes the maximum duration of a pulse cycle that must be considered.

A LowerRateLimit (LRL) parameter indicates the minimum period of a pulse below which the heart must be stimulated by the pacemaker. This means that a cycle must not last longer than the interval indicated by the LRL parameter. It is expressed thus as follows:

REQUIREMENT 8.2.– The period of a cycle must not be higher than the LRL interval.

8.3. Model-checking of the requirements

To establish the verification of a set of requirements on a model that must be validated, we need to use a model that can be simulated, as well as of requirements that are formulated, for example, by logical formulas (Linear Temporal Logic (LTL) and Computation Tree Logic (CTL)) or by automatic observers (see Chapter 1). We must also model the behaviors of the environment that interacts with the model we must validate. This environment corresponds to different use cases of the system that we must consider for requirement verification.

8.3.1. *Use case*

To showcase the analysis carried out on the model, in this chapter, we will concern ourselves with the analysis of relevant requirements when the pacemaker is in nominal mode, that is in active mode. In addition, we will consider this particular context of use. Given the specs provided, the relevant parameters that we must take into account are the choice among the seven modes ($mode$) and the different values Atrial Refractory Period (ARP), Ventricular Refractory Period (VRP), $AVDelay$ and LRL.

The DCM interacts with the pacemaker by sending it the values for the different parameters, and their potential values can be multiple. In this context, the DCM sends out the values for the five parameters: $mode$, ARP, VRP, $AVDelay$ and LRL in any order. In our modeling, the sending out of the values for the DCM to the pacemaker is done by communicating a $chgParam$ message to the destination of the $PGController$ process. This message has two arguments: the name of the parameter that must be changed and the new value. For instance, in order to put the pacemaker in mode VOO, the message sent is $chgParam$ $(mode, \ VOO)$. Figure 8.3 shows an interaction that implements the five messages for the mentioned parameters.

The different values that these parameters can take and that we have considered in our Fiacre modeling are as follows:

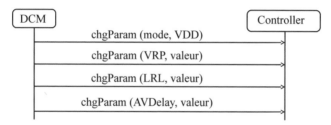

Figure 8.3. *An example of an interaction scenario (in mode VDD) between DCM and PGController*

– parameter mode: seven possible modes: VOO, AOO, VVI, AAI, VVT, AAT, VDD;

– ARP parameter: solely in modes AAI and AAT with 36 possible values in each of them (from 150 to 500 ms at a pace of 10 ms);

– VRP parameter: solely in modes VVI, VVT and VDD with 36 possible values in each of them (from 150 to 500 ms at a pace of 10 ms);

– AVDelay parameter: solely in the VDD mode with 24 possible values (from 70 to 300 ms at a pace of 10 ms);

– LRL parameter: in all the modes, except AOO and VOO, with 62 possible values (from 30 to 175 ppm at a pace of 5 ppm up to 50 ppm, then a 1 ppm pace up to 90 ppm and then a 5 ppm pace up to 175 ppm).

Therefore, the number of values that the parameters can take on depends on the modes. This is presented in the Table 8.1. The total number of possible combinations, that is of possible instances of the scenario presented in Figure 8.3, is thus 62,498.

Modes	AOO	VOO	AAI	AAT	VVI	VVT	VDD
AVDelay values	–	–	–	–	–	–	24
ARP values	–	–	36	36	–	–	–
VRP values	–	–	–	–	36	36	36
LRL values	–	–	62	62	62	62	62
Number of combinations	1	1	2,232	2,232	2,232	2,232	53,568

Table 8.1. *Number of different values for the parameters*

8.3.2. *Properties*

Requirement 8.1 can be expressed under the form of an invariant of the system. In VDD mode, the Cardiologist process enters the state *TrackedPacing_WaitingVPulse* when it receives an auricular signal and waits for a ventricular signal. The variable timeCount of the Cardiologist then measures in units the time that has passed in that state. This allows us to transform the expression of the requirement in the following invariant:

INVARIANT 8.1.– The state `TrackedPacing _ WaitingVPulse` of the `Cardiologist` process implies that the value of the `timeCount` is lower than or equal to `AVDelay`.

This invariant is written in SELT linear logic for the *model-checker* TINA SELT:

```
[] ( cardio_1_sTrackedPacing__WaitingVPulse
    => {cardio_1_vtheMode.AVDelay} >= cardio_1_vtimeCount
   );
```

Requirement 8.2 could equally be expressed in the form of an invariant of the model. However, this would require that we adapt the model by adding extra variables that could be referenced by the invariant. Without these variables, the time that has passed in the current cycle would not be known by our system. We will thus choose an alternative that consists of encoding the requirement using an observing automaton (*Obs 2*) [HAL 93], illustrated in Figure 8.4. Let us explain its principle.

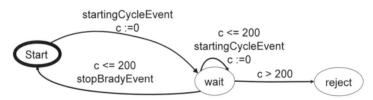

Figure 8.4. *Obs 2: timed monitor that encodes requirement 8.2*

An observing automaton is an automaton that is sensitive to events taking place throughout the exploration of the mode and of the environment, events such as sending and receiving messages, changing the state of the processes and the variable values. With each execution of a transition in the model or the environment, the observer executes a transition (if it exists) that corresponds to the event that just took place. The observer has an error state *reject*. Access to the *reject* state means that the property encoded by the automaton has been falsified. An accessibility analysis consists of researching possible scenarios that have been observed during the exploration of the system model and its environment, leading to the *reject* state of the observer. Several observers can thus be defined by the same exploration of the model. Regarding the observers, we can thus encode certainty-type properties as well as limited vitality-type properties. The coding interest on the part of the observers is to be able to express the properties that are more difficult to express in temporal logic such as LTL or CTL. The observers cannot be managed in the current version of TINA SELT, so we will use the OBP Explorer tool to check them.

To meet requirement 8.2, to be able to measure the time that separates the two beginnings of a cycle, we will time the observer. The timing must be done with the help of a clock whose value increases in synch with the time of the simulation of the model. The clock can be reset to zero when a transition has been executed. The clocks of the observing automata are being managed within the OBP Explorer tool in accordance with the TTS semantic.

In the example in Figure 8.4 that encodes requirement 8.2, the observer goes into *wait* mode as soon as the `startingCycleEvent` occurs. This event is defined starting from a predicate `startingCycle` that we will specify as follows:

$$event\ startingCycleEvent\ is\ \{\ startingCycle\ becomes\ true\}$$

This specification corresponds to the description format of the predicate that we use in CDL language, which is described in section 8.4.3. The predicate `startingCycle` is thus defined starting from the *TrackedPacing_ StartingPulse, AsynchronousPacing, InhibitedPacing_StartingPulse* and *TriggeredPacing_WaitingPulse* states of the `Cardiologist` automaton and the `timeCount` variable:

$$
\begin{aligned}
predicate\quad & startingCycle\ is\ \{ \\
& (\ \{cardiologist\}1@TrackedPacing\_StartingPulse\ or \\
& \{cardiologist\}1@AsynchronousPacing\ or \\
& \{cardiologist\}1@InhibitedPacing\_StartingPulse\ or \\
& \{cardiologist\}1@TriggeredPacing\_WaitingPulse\ ) \\
& and\ \{cardiologist\}1:timeCount\ =\ 0\}
\end{aligned}
$$

A clock c is associated with the observer. The `stopBradyEvent` is detected when the stimulation is paused, that is when the message *StopBrady* sent by the `PGController` is received. This is specified thus:

$$
\begin{aligned}
event\quad & stopBradyEvent\ is \\
& \{\ receive\ StopBrady\ from\{PGController\}1\ to\ \{cardiologist\}1\}
\end{aligned}
$$

When this event is detected, the observer returns to its initial state, waiting for a new cycle to begin. The *reject* state of the observer is reached if the value of the clock goes higher than the *LRL* value. The clock is reset to zero with each detection of `startingCycleEvent`. The value of the clock c thus represents the time that has passed within the same cycle.

8.3.3. *Property check*

To verify the properties on the model, the latter must be composed with the use case that describe the environment of the system, then simulated and explored by a checking tool. The exploration generates a transition system (TTS). This represents all the behaviors of the model in its environment under the form of a graph of configurations and transitions. On this TTS, we may carry out a property check, either (Figure 8.5(a)), by applying, as in the case of the TINA SELT tool *model-checking* algorithms on logical formulas or as in the case of the OBP Explorer tool (Figure 8.5(b)), an accessibility analysis of *reject* states of the observers. The difficulty with this technique is the production of the TTS that can have a large size, larger than the size of the available memory (combinatorial explosion). If the available memory is insufficient, the verification is not usually completed.

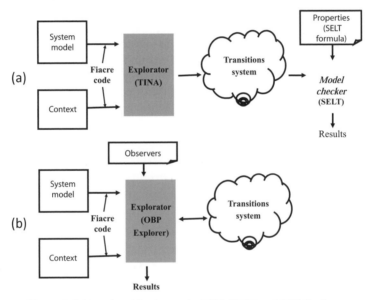

Figure 8.5. *Tested verification tools: TINA SELT and OBP Explorer*

To be able to check the two properties mentioned, we must integrate, in the Fiacre model, the behavior of the process DCM in order to simulate sending the values for the five parameters to the PGController process. We will use both the tools mentioned previously[4]: TINA SELT for checking invariant 8.1 and the OBP Explorer tool for the verification of observer 2 in Figure 8.4. Both these tools differ in the moment where the actual verification takes place. In the case of the TINA SELT chain, the TINA explorer explores the system composed with its environment, generates a transition

---

4 In both cases, the verifications are carried out on a PC, with two gigabytes of memory.

system (the graph of global behaviors), and then in a second stage the SELT *model-checker* checks the property on that graph. Conversely, the OBP Explorer generates the behavior graph by taking into account the properties that must be verified (coded as observers) and carries out the verification "on the fly", that is at the same time as the behavior graph is being generated. The verification, in this case, is reduced to searching for the *reject* states in the graph.

The Table 8.2 presents the results of the verification of invariant 8.1, with the TINA SELT tool, for different numbers of possible combinations of value exchange between the five parameters.

The verification must be carried out on the entire set of combinations (53,568 for the VDD mode). But the verification does not end because of a combinatorial explosion. We will thus test several combinations in order to evaluate the complexity that can be processed with our tools. We note there has been an explosion for 2,592 combinations. We have only been able to explore less than 4% of the space of these combinations, and thus only 4% of the space of the behaviors. Therefore, we cannot deduce a result for our verification. Table 8.2 gives us, for each number of possible combinations, the number of configurations (states of the transition system) and transitions explored as well as of the *model-checking* time for this *VDD* mode.

Number of combinations	Number of explored configurations	Number of explored transitions	Time (sec)
288	321,170	748,218	12
490	541,192	1,260,682	22
768	839,834	1,956,308	33
1,134	1,233,560	2,873,322	52
1,600	1,734,954	4,041,026	219
2,592	Explosion	–	–

Table 8.2. *Verification with TINA SELT of the invariant 8.1*

Table 8.3 presents the results for the verification of requirement 8.2, for all the modes, with the OBP Explorer tool. All of the modes involve the application of a set of 62,498 combinations. The combinatorial explosion takes place from 352 combinations onwards (or less than 0.6% of the space of the combinations), which stops us from being able to check the properties beyond this complexity.

Let us note that here the explosion takes place for a number of combinations (352) that is much lower than in the case of the check for invariant 8.1 with TINA (i.e. 2,592). This is explained by the fact that the introduction of an observer in the exploration and accessibility analysis adds a component to the composition of the model, which in turn increases the size of the generated transition system.

Number of combinations	Number of explored configurations	Number of explored transitions	Time (sec)
7	40,455	48,781	2
42	299,740	364,391	10
110	744,224	909,908	25
226	1,783,438	2,238,305	64
352	Explosion	–	–

Table 8.3. *Verification with an OBP Explorer of the observer that corresponds to requirement 8.2*

8.3.4. *First assessment*

As a result of these checks, we note that the combinatorial explosion of the size of the TTS's stops us from verifying the requirements beyond a complexity that is quickly reached. The number of reachable configurations in the model explored quickly becomes too large to be memorized because these configurations are generated on the set of the model's behaviors. And this, despite the environment of the system and its encoding in a Fiacre model composed with the model of the system. A first explanation comes from the phenomenon already observed when we consider the observers in the exploration of the system model and its environment using an OBP Explorer tool. Considering a new actor increases the size of the space of generated states and thus reinforces the combinatorial explosion.

Generalizing from this observation, we may note that, if considering the environment in Fiacre models is indeed necessary, then considering all the behaviors of the environment at once can only lead to a more rapid explosion. To overcome this difficulty, and to check the properties on a number of more significant combinations, we will implement a technique whose goal is to limit the complexity of the TTS's by breaking down the environment in several use cases.

The objective is to verify the requirements, not only in a single exploration, but on several explorations. Each exploration corresponds to a subset of behaviors of the DCM in such a way that the exploration becomes possible on state spaces that are smaller and smaller in size. We will describe this technique in the following section.

8.4. Context exploitation

When we do the classical implementation of the *model-checking*, the model explored includes the use scenarios. The use scenarios describe the interactions between the components of the environment and the model. Considering the model with the behavior of the environment leads to the necessity of exploring a state space that is, most of the times, very large (Figure 8.5).

Here, we choose to specify the environment not only as a single global process, but also as a set of explicit and separate scenarios called *contexts*. In this approach, the environment is described as the union between all of the contexts. Each context allows us to activate subsets of behaviors of the model. However, the requirements that we must verify are certainty or accessibility properties, and they are satisfied if and only if they are satisfied in all the contexts taken separately [ROG 06]. This enables us, with the help of *model-checking*, the need to no longer explore a "large" state space but several tinier spaces (as many as there are contexts) (Figure 8.6). This "divide and conquer" approach also allows us to carry out the verifications on systems of significant size. We specify here the principle implemented with the OBP tool.

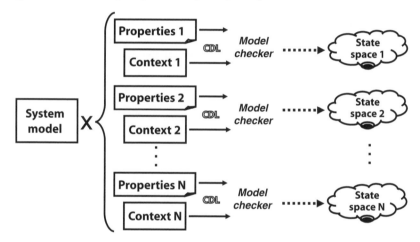

Figure 8.6. *Exploration with identification of separate contexts*

8.4.1. *Identifying the context scenarios*

To identify the contexts and to be able to formally specify them, the user relies on the knowledge they have of the environment of the system. They generally correspond to use modes of the modeled component. In the context of reactive embedded systems, the environment of each component of a system is often well known. It is thus more efficient to identify this environment than to seek to reduce the space of the model configurations of the system that is to be explored.

The objective is to have a description of the subsets of behaviors displayed by the actors in the environment (Context$_i$, i $\in$ [1..N] in Figure 8.6) and of the subsets of properties associated with these behaviors. This identification of the contexts can allow us to bypass the explosion when exploring the model.

In order for this approach to be well grounded, the process of system development must include a stage where we specify the environment, thus allowing us to accurately identify the sets of concluded and completed behaviors.

The strong hypothesis underlying the implementation of this methodological process is that the designer is capable of identifying all of the possible interactions between the system and its environment. We will also assume that each context expressed in the beginning is also finished, namely that the scenarios described by this context do not have infinite iterative behaviors.

In the field of embedded systems, in particular, we base this hypothesis on the fact that a software component designer must accurately and fully comprehend the perimeter of its use (i.e. constraints and conditions) to develop it correctly. It will be necessary to carry out a formal study of the validity of this hypothesis according to targeted applications. We will not approach this aspect that requires a specific methodological work to be carried out. However, in our current approach, we must include the infinite behaviors of certain entities of the environment in our system model.

8.4.2. *Automatic partitioning of the context graphs*

When the restriction of the behaviors of a model does not suffice, as previously described, namely when one of the Context$_i$ leads to an exploding state space, we implement a second lever meant to reduce the space of these states. Each context that leads to an explosion is automatically and recursively partitioned in a set of ever more reduced subcontexts (Figure 8.7).

To do this, we will implement a recursive partitioning (or splitting) algorithm in our OBP tool. Figure 8.7 illustrates the `explore_mc()` function for the exploration of a *model*, with a *context* and the verification for a set of properties *pty*. The context is represented by an acyclic graph. This graph is composed of the model throughout the exploration. In the case of explosion, this context is automatically split into several sub-graphs (including parameter d, which specifies the depth of the graph where the splitting starts). This recursive processing is carried out until all of the explorations have been implemented without leading to an explosion.

In fact, for every Context$_i$, we transform a global verification in tinier K_i verifications, where K_i is the number of subcontexts obtained after partitioning the Context$_i$. In the end, for the set of contexts we must transform the N verifications (as many verifications as there are contexts) into $N' = \sum_{i=1}^{N} K_i$ smaller verifications. We must thus note that the implemented partitioning technique observes the following principle: for a given context, the union of the executions – described by the set of sub contexts generated by the partitioning of the context – includes the executions described by this initial context. The properties are thus preserved by the partitioning of the context as shown in [ROG 06].

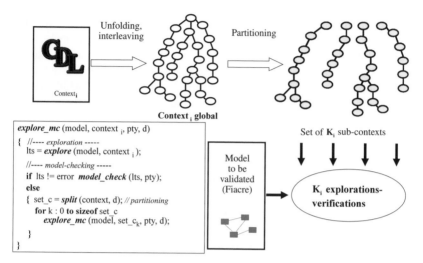

Figure 8.7. *Partitioning a context and checking each partition*

8.4.3. *CDL language*

CDL [DHA 09] is a domain-specific language (DSL) drawing from a *use case chart* by [WHI 06] based on activity diagrams and sequence diagrams. The objective of CDL is to allow the formal modeling of the contexts of the systems (i.e. the Contexts$_i$ mentioned in the previous sections). This formalism enables us to describe the behavior of several entities (called *actors*) that make up the environment. They are executed at the same time and interact with the model of the system.

Equipped with a graphic and textual syntax, a CDL model describes, on the one hand, a scenario with the activity diagrams and the sequence diagrams. On the other hand, it describes the properties we must verify by relying on property defining patterns. A CDL meta-model was defined as well as a syntax and a formal semantic that are described [DHA 11a] in terms of traces[5], drawing from the works of [HAU 05] and [WHI 06].

A CDL model is structured in a hierarchical fashion. A first level specifies the set of actors that make up the environment of the system and that evolve at the same time. At this level, a CDL context can be represented (in a simplified manner) by a

---

5 For the detailed syntax, see also [DHA 11b] available on www.obpcdl.org

construction of type $A_1 \parallel A_2 \parallel \ldots \parallel A_n$ where each A_i represents an actor of the environment. Each A_i is then detailed in an activity diagram, namely as a composition of Message Sequence Chart (MSC) or activity sub-diagrams. The possible combinations at this level being the interleaving between two or more branches (i.e. the sequence) or the choice between two or more branches (i.e. the alternative). Then, each MSC is described as a simplified sequence diagram of UML2.0 type [ITU 96], which describes the interactions between the actor and the system. Formally, a CDL model can be considered as a set of MSCs composed between them with the help of two operators: the sequence (seq), the parallel (par) and the alternative (alt).

When we compile a CDL model, the diagrams that correspond to each actor are laid out (examined from the angle of each finished loop) then interleaved according to Fiacre semantic and at the same time with UML-MARTE semantic. The interleaving of the set of MSCs, describing the behavior of the context, generates a graph that represents all of the executions of the actors in the environment considered. This graph is then split in such a way as to generate a set of sub-graphs that correspond to subcontexts, as mentioned in section 8.4.2. When the observer explores the environment, each sub-graph is composed (Figure 8.7) of the model that needs to be validated, and the properties are checked against the result of this composition.

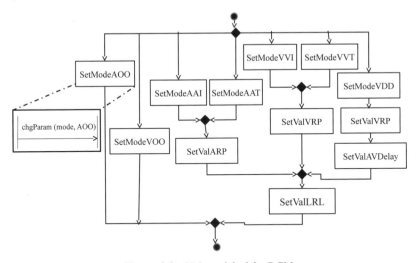

Figure 8.8. *CDL model of the DCM*

Figure 8.8 graphically represents a CDL model of the DCM (which is the only actor in the pacemaker's environment). In this context, the DCM can run one of the seven MSCs *SetModeX* (the operation "alternative" is graphically represented by a diamond). Each MSC *SetModeX*, where X is 'AOO', 'VOO', 'AAI', 'ATT', 'VVI', 'VVT' or 'VDD', models the sending of an instruction of changing the mode of the DCM toward the pacemaker. For example, $SetModeAOO$ is an MSC containing the unique interaction "chgParam (mode, AOO)" between the DCM and the PGController (see Figure 8.3). This mode changing having been executed, the DCM continues by executing zero, one, two, or three MSCs $SetValX$, depending on the branch caught in the first alternative. Each $SetValX$ is an MSC that models all the possible alternatives for the allocation of a value at the parameter X, where X is "ARP", "VRP", "AV Delay" or "LRL". For example, the MSC $setValARP$ (Figure 8.9) models an alternative between 36 possible allocations of the parameter ARP. The CDL model in Figure 8.8 thus represents the set of 62,498 possible combinations at the input of the pacemaker.

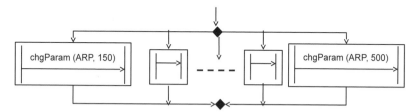

Figure 8.9. *SetValARP: value exchange for the ARP parameter*

Finally, let us note that CDL thus enables us to specify the properties starting from definition patterns that define the properties in such a way as to help the user formalize them. We will not, however, discuss this aspect in this chapter.

8.4.4. *CDL model exploitation in a model-checker*

The OBP tool takes as an input the CDL models, in order to convert them in two ways, as showcased in Figure 8.10. Let OBP translate the CDL diagrams in Fiacre programs so that they are later submitted to TINA, and let OBP generate the data for the OBP Explorer, in itself integrated in OBP.

The Fiacre programs, or the data generated by OBP, describe in both cases a set of acyclic context graphs. They represent the set of possible interactions between the environment and the model. To validate the latter, it is both necessary and sufficient to compose them with each graph [ROG 06]. The properties must thus be verified against the result of each of these compositions. In the case of TINA, the SELT formulas are verified by *model-checking*. In the case of OBP Explorer, an accessibility analysis is carried out on the result of the composition between a generated graph, a set of observers and the model. In these two cases, OBP recovers the results provided by SELT or by OBP Explorer and formats them to make them comprehensible for the user. The splitting of the context by the OBP in a set of graphs allows us to reach, within the composition, limited-sized transition systems, thus, ensuring the accessibility analysis in the case of an OBP Explorer or the *model-checking* in the case of TINA.

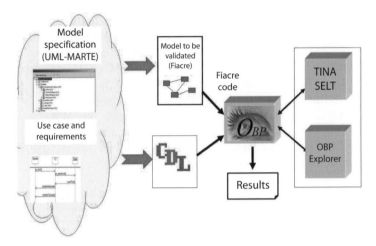

Figure 8.10. *Transformation of a CDL model with OBP*

8.4.5. *Description of a CDL context*

For the case study in question, the CDL context describes the interactions between the DCM and the PGController process. It thus includes only one CDL actor. The interactions are described as alternatives and sequences that model all of the envisaged scenarios, namely the combinations of values provided to the PGController. One of these scenarios, as illustrated in Figure 8.8, describes sending out the mode value to the PGController process and then sending a value for each of the parameters ARP, VRP, LRL and $AVDelay$. The textual version of the set of scenarios illustrated by Figure 8.8 is named Listing 8.3. The operator "[]" specifies the alternative and ";" specifies the sequence. The $allModes$ scenario references $refractoryPeriodProperty$. We are concerned with the property that corresponds with the observer *Obs 2* to be checked.

```
cdl allModes is {
properties refractoryPeriodProperty;
{
    { set_mode_AOO [] set_mode_VOO }
    []
    {
      {
          { { set_mode_AAI [] set_mode_AAT }; set_val_ARP }
          []
          { { set_mode_VVI [] set_mode_VVT }; set_val_VRP }
          []
          { set_mode_VDD; set_val_VRP; set_val_AVDelay }
      };
      set_val_LRL
    }
}
}
```

Listing 8.3. *CDL model (text version) for all the modes* where:

– set_val_ARP equivalent to: $setARP\_150$ [] ... [] $setARP\_500$;

– set_val_VRP equivalent to: $setVRP\_150$ [] ... [] $setVRP\_500$;

– $set\_val\_AVDelay$ equivalent to: $setAVDelay\_70$ [] ... [] $setAVDelay\_300$;

– set_val_LRL equivalent to: $setLRL\_30$ [] ... [] $setLRL\_175$.

8.4.6. Results

We present the results of the verification of the properties by implementing the CDL contexts and the partitioning technique of the context graphs.

Table 8.4 presents the results for the verification of requirement 8.1 for the mode VDD with the TINA SELT tool with the exploitation of the contexts. Contrary to the experimentation shown in section 8.3.3, the verification ends in approximately half an hour including the full check of the DCM, namely offering the 53, 568 possible combinations of the VDD mode (last line of the table). The verification shows that the requirement 8.1 is satisfied by the model of the pacemaker.

Number of combinations	Number of generated sub contexts	(Cumulated) Number of explored configurations	(Cumulated) Number of explored transitions	Exploration time (sec)
1,600	10	1,735,440	4,042,232	72
2,592	12	2,807,562	6,538,731	125
6,144	16	6,598,944	15,367,920	268
12,000	20	12,790,310	29,786,725	508
20,736	24	22,059,672	51,371,796	855
32,928	588	34,932,422	81,367,573	1,499
47,616	768	50,791,728	118,294,920	2,017
53,568	864	57,148,944	133,100,760	2,270

Table 8.4. *Verification of the invariant 8.1 (mode VDD) with TINA SELT and the exploitation of contexts*

To illustrate the evolution of the verification time, we have also tested the verification method on the subsets of DCM ranging from 1,600 possible combinations (from this context configuration, the classical method in section 8.3.3 arises) to the totality of combinations. We can then note that the verification time evolves fairly exponentially, by remaining reasonable (maximum half an hour) for a space of really large states (up to 57 million configurations).

Table 8.5 showcases the results for the verification of the observer *Obs 2* corresponding to requirement 8.2, for all the modes, with the exploitation of the CDL model of the DCM by OBP. Here as well, we note that the verification finishes in approximately 6 h, including the complete model of the DCM offering 62,498 possible combinations (last line of the table). The space of the explored states contains around 535 million configurations. As in the previous case, the time evolution of the verification is shown by a succession of experiments carried out on ever larger contexts (comprising 352 combinations in total). Here as well, the time evolution is exponential, but remains feasible in practice (approximately 6 h).

Number of combinations	Number of generated sub contexts	(Cumulated) Number of explored configurations	(Cumulated) Number of explored transitions	Time of exploration (sec)
352	7	2,680,017	3,362,387	92
578	12	4,519,385	5,729,766	238
884	13	7,033,893	8,987,094	241
1,282	14	10,179,272	13,069,412	437
3,746	72	30,869,430	40,072,535	1,686
8,194	228	67,893,326	88,639,720	3,963
15,202	344	126,267,775	16,5504,813	6,355
25,346	484	21,2075,376	27,8725,276	9,679
39,202	648	329,522,688	434,009,683	14,226
55,554	636	476,083,965	628,001,017	19,779
62,498	940	535,871,149	706,896,539	22,238

Table 8.5. *Observing requirement 8.2 with OBP Explorer and the exploitation of contexts*

Let us note once again that the size of the generated transition system is more important than in the case of exploration with TINA. For the same number of combinations, its size is here 10 times higher. As explained in section 8.3.3, introducing the observer throughout the exploration and the accessibility analysis increases the complexity of the generated transition system.

8.5. Assessment

The identification of CDL contexts and the partitioning technique allow us to restrict the execution of models and enables us to carry out the explorations till the end. In the case of TINA or of the analyzer internal to the OBP, the exploration of all the behaviors is impossible without implementing a context model as well as the partitioning technique of this model. We have shown that the application of the set of 62,498 combinations of the values of the parameters associated with the chosen modes can lead to a verification of the two requirements considered. In addition, the input of CDL models offers a formal framework for describing the requirements that can be automatically translated into observing automata.

In industrial applications, the description of the contexts that interact with the model to be validated is often informal, and sometimes incomplete. The approach allows the user to formalize this environment and to specify, in a set of CDL models, the cases in which the component developed will be used. This formalization can only improve the design even if the user does not exploit it by using it for formal analysis. This formalization is based on activity diagrams and sequence diagrams, which are more easily accessible for an engineer. Conceptually, the CDL principles can thus be implanted without further, more standardized formalisms, such as activity diagrams and UML sequences.

In our illustration, we have considered the DCM to be the sole actor in the environment. Because we have a unique actor in this case, the scope of the automatic partitioning may seem limited. However, this is not the case as the environment is made up of several actors. OBP generates, from the behavior of the actors, a graph of the set of actors' behaviors considering the interleaving of these behaviors. This graph is then split using the technique previously described in this chapter.

We could carry out verifications on more complex models, with a much larger number of combinations. The exploration as well as the actual check would last significantly longer. To speed up the verification, we should increase the performance of the equipment by, first, increasing its memory so that we can limit the operations of context partitioning and, second, by carrying out the processing of contexts on a network of machines [DHA 11a].

8.6. Conclusion

The objective of this chapter is to present a verification technique (by *model-cheking*) to check the properties on the UML-MARTE model described in Chapter 7.

The approach described seeks to reduce the space of possible behaviors by considering an explicit model of the environment as a union of contexts. This model is described thanks to CDL language that allows us to describe the interactions between the environment and the formalized model that needs to be validated.

A context thus formalized can be exploited by the OBP tool and split into subcontexts that are composed of the model that needs to be validated. This partitioning allows us to reduce, with every verification, the number of behaviors of the model to be explored, thus also reducing the combinatorial explosion. Throughout the explorations, we implement and verify the observers. We have illustrated this technique for a set of interactions between the DCM and the pacemaker controller. This technique is implemented on the TINA *model-checker* and the OBP Explorer analyzer, which is internal to the OBP.

8.7. Bibliography

[ALU 97] ALUR R., BRAYTON R.K., HENZINGER T.A., *et al.*, "Partial-order reduction in symbolic state space exploration", in *Computer Aided Verification*, volume 1254 of LNCS, pp. 340–351, 1997.

[BER 04] BERTHOMIEU B., RIBET P.-O., VERDANAT F., "The tool TINA – construction of abstract state spaces for Petri nets and time Petri nets", *International Journal of Production Research*, vol. 42, no. 14, pp. 2741–2756, July 2004.

[BOS 05] BOSNACKI D., HOLZMANN G.J., "Improving spin's partial-order reduction for breadth-first search", *SPIN*, vol. 3639, pp. 91–105, 2005.

[CLA 86] CLARKE E., EMERSON E., SISTLA A., "Automatic verification of finite-state concurrent systems using temporal logic specifications", *ACM Transactions on Programming Languages and Systems*, vol. 8, no. 2, pp. 244–263, 1986.

[DHA 09] DHAUSSY P., PILLAIN P.-Y., CREFF S., *et al.*, "Evaluating context descriptions and property definition patterns for software formal validation", *12th IEEE/ACM Conference Model Driven Engineering Languages and Systems (Models'09)*, Springer-Verlag, vol. 5795 of LNCS, pp. 438–452, 2009.

[DHA 11a] DHAUSSY P., BONIOL F., ROGER J.-C., "Reducing state explosion with context modeling for model-checking", *13th IEEE International High Assurance Systems Engineering Symposium (Hase'11)*, Boca Raton, FL, 2011.

[DHA 11b] DHAUSSY P., ROGER J.-C., CDL (context description language): Syntaxe et sémantique, Report, ENSTA-Britain, 2011.

[EME 97] EMERSON E., JHA S., PELED D., "Combining partial order and symmetry reductions", *Tools and Algorithms for the Construction and Analysis of Systems*, Springer Verlag, Enschede, Netherlands, LNCS 1217, pp. 19–34, 1997.

[FAR 08] FARAIL P., GAUFILLET P., PERES F., *et al.*, "FIACRE: an intermediate language for model verification in the TOPCASED environment", *European Congress on Embedded Real-Time Software (ERTS)*, SEE, Toulouse, January 2008.

[FER 96] FERNANDEZ J.-C., GARAVEL H., KERBRAT A., *et al.*, "CADP: a protocol validation and verification toolbox", *Proceedings of the 8th International Conference on Computer Aided Verification CAV '96*, Springer-Verlag, London, UK, pp. 437–440, 1996.

[GOD 96] GODEFROID P., PELED D., STASKAUSKAS M.G., "Using partial-order methods in the formal validation of industrial concurrent programs", in *Proceedings of the 1996 International Symposium on Software Testing and Analysis (ISSTA '96)*, ZEIL S.J., and TRACZ W. (eds.), ACM Press, New York, San Diego, CA, pp. 261–269, 8–10 January 1996.

[HAL 93] HALBWACHS N., LAGNIER F., RAYMOND P., "Synchronous observers and the verification of reactive systems", in NIVAT M., RATTRAY C., RUS T., SCOLLO G., (eds), *Third International Conference on Algebraic Methodology and Software Technology (AMAST'93)*, Springer Verlag, Twente, pp. 83–96, June 1993.

[HAU 05] HAUGEN O., HUSA K.E., RUNDE R.K. *et al.*, "STAIRS towards formal design with sequence diagrams.", *Software and System Modeling*, vol. 4, no. 4, pp. 355–357, 2005.

[HEN 91] HENZINGER T., MANNA Z., PNUELI A., "Timed transition systems", in *Real-Time: Theory in Practice, Proceedings of REX Workshop (REX'91)*, Mook, The Netherlands, 3–7 June 1991, LNCS 600, Springer-Verlag, pp. 226–251, 1992.

[HOL 97] HOLZMANN G., "The model checker SPIN", *Software Engineering*, vol. 23, no. 5, pp. 279–295, 1997.

[ITU 96] ITU, Message sequence chart (MSC), ITU-T Recommendation Z.120, Geneva, 1996.

[LAR 97] LARSEN K.G., PETTERSSON P., YI W., "UPPAAL in a nutshell", *International Journal on Software Tools for Technology Transfer*, vol. 1, no. 1–2, pp. 134–152, 1997.

[MCM 92] MCMILLAN K.L., PROBST D.K., "A technique of state space search based on unfolding", *Formal Methods in System Design*, vol. 6 no. 1. pp. 45–65, January 1995.

[PAR 06] PARK S., KWON G., "Avoidance of state explosion using dependency analysis in model checking control flow model", in *Proceedings of the 5th International Conference on Computational Science and its Applications (ICCSA '06)*,vol. 3984 of Lecture Notes in Computer Science, pp. 905–911, 2006.

[PEL 98] PELED D., "Ten years of partial order reduction", *Proceedings of the 10th International Conference on Computer Aided Verification (CAV '98)*, Springer-Verlag, pp. 17–28, 1998.

[QUE 82] QUEILLE J.-P., SIFAKIS J., "Specification and verification of concurrent systems in CESAR", *Proceedings of the 5th Colloquium on International Symposium on Programming*, Springer-Verlag, London, UK, pp. 337–351, 1982.

[ROG 06] ROGER J.-C., Exploitation de contextes et d'observateurs pour la validation formelle de modèles, PhD thesis, ENSIETA, University of Rennes I, December 2006.

[VAL 91] VALMARI A., "Stubborn sets for reduced state space generation", *Proceedings of the 10th International Conference on Applications and Theory of Petri Nets*, Springer-Verlag, London, UK, pp. 491–515, 1991.

[WHI 06] WHITTLE J., "Specifying precise use cases with use case charts", *9th IEEE/ACM Conf. Model Driven Engineering Languages and Systems (MoDELS'06)*, Satellite Events, Genova, Italy, pp. 290–301, 1–6 October 2006.

Chapter 9

Model-Based Deployment
and Code Generation

9.1. Introduction

The objective of this chapter is to describe the process of generating a set of executable binary files from the design model of the pacemaker, discussed in Chapter 8. This design model is based on executable components, which specify the structure, the internal behavior and the interactions between these components.

The structure is described by the decomposition of the system into components and the interactions between them. The interactions are modeled by placing the components in relation with each other via connectors between dedicated interaction points of a component called ports. The well-formedness of the interaction model depends on the compatibility between the interrelating ports. The component's behavior is specified by the algorithmic aspect, which describes the services offered by the component (i.e. its operations), and the control aspect, which describes the orchestration of the service calls and the different modes in which the component functions.

The component's behavior is described algorithmically, which includes the services offered by the component (i.e. its operations) and from the control perspective, which includes the orchestration of the service calls and functional modes of the component. The control aspect of each component is modeled by a state machine. The algorithmic aspect is specified by an action language.

Chapter written by Chokri MRAIDHA, Ansgar RADERMACHER and Sébastien GÉRARD.

To this model of executable components, we add a description of the execution platform and a deployment description on this platform. The description of the platform allows us to identify the number of execution nodes. This information is used to allocate components on these nodes. The allocation allows us to identify the number of executable binaries that we must generate and the content of each binary.

The model of executable components is made up of components, ports, connectors, state machines and activities that must be translated in a third-generation programming language to produce executable binaries (while still preserving the execution semantic of these concepts). It is important to note that some of these modeling concepts have no equivalent in a programming language. For example, no third-generation programming language offers the concept of the state machine. Therefore, defining the projection of state machines to a programming language requires the implementation of patterns called "code generation patterns".

To limit the complexity of code generation patterns, it is important to choose a target programming language that offers concepts that are close to the modeling language – in our case, Unified Modeling Language (UML). From the point of view of its history and of the concepts offered by UML, object-oriented programming languages turn out to be good target languages for our purposes. They offer the concept of class on which the conceptual definition of a Modeling and Analysis for Real-Time Embedded systems (MARTE) component is based (see Chapter 7). Therefore, in this chapter, we will use C++ language as target programming language for generating code.

Figure 9.1 describes the main stages for generating an executable binary starting from an executable component-based model, a platform description model and a deployment description model.

The input models must be well formed in order to enable their automatic processing. Therefore, the first stage of the process of generating the executables, described in Figure 9.1, consists of validating these models. A set of rules verifies that these models are coherent for an automatic processing by the generator chain. If that is not the case, the errors are indicated and returned to the designer, to modify or complete the input models before redoing the validation process.

The second stage consists of generating an implementation model on the basis of valid input models. This implementation model is a component-based model making implementation mechanisms explicit. Here, we apply implementation patterns that regard communication and behavior. When this stage begins, a global implementation model common to all execution nodes is generated.

This model is provided at the beginning of the stage where we produce code in an object-oriented programming language for each execution node. In the last stage, this code is compiled in order to produce an executable binary for each execution node of the platform.

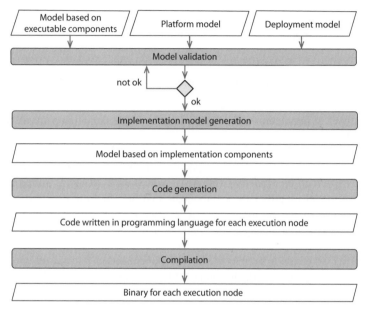

Figure 9.1. *Global view of the approach*

This chapter is organized as follows. Section 9.2 provides a description of input models on the basis of the model of the pacemaker. Section 9.3 describes the generation of the component-based implementation model and section 9.4 provides the details for code generation by describing the activities and the intermediary models for this stage. In section 9.5, we introduce a tool support for this approach.

9.2. Input models

This section presents the input models of the process, namely the component-based model, the platform model and the deployment model.

9.2.1. *Description of the executable component-based model*

The MARTE component-based model that we have used corresponds to the model presented in section 8.3. This model describes the components (along with their internal behavior), the interaction points, (ports) characterized by the data they convey or the services that they provide or require, as well as the topology of the interconnections between components, which is specified by the connectors between the ports.

The interaction model, namely the type of communications, is specified by this component-based model. The types of communications are:

– synchronous or asynchronous for the operation calls (*clientServerPort*);

– *push* or *pull* for the data communications (*flowPort*).

These types of interaction will be carried out by implementation patterns described in section 9.3.

The component-based model also specifies an abstract concurrency model using *High-level Application Modeling* (HLAM), which is a MARTE subprofile. This concurrency model identifies the components that benefit from execution resources (RtUnit) and the shared components (PpUnit) that do not have execution resources, but rather resources that manage their concurrent accesses.

Figure 9.2 provides a sample of the component-based pacemaker model. This system is made up of two main components: the DeviceControlMonitor and the PGController. These two components communicate with each other via ClientServer ports.

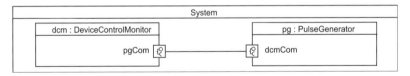

Figure 9.2. *Component-based model of the pacemaker system*

9.2.2. *Description of the platform model*

Somewhat similarly to a software architecture, a hardware architecture can be described as a composition of elements. A class, called "HWArchitecture", globally represents the platform. The attributes of this class represent the nodes. Each node is coupled with a class that has the same properties as that node. Each node may have an internal structure, which makes the hierarchical structure easier to decompose.

Figure 9.3 describes the definition of the platform for the case study. The components of this platform are modeled and characterized by using the concepts of the *Hardware Resource Modeling* (HRM) MARTE subprofile. The platform is made up of two execution nodes (PaceMakerNode and DCMControlNode), both stereotyped as HwProcessors, and a communication bus, stereotyped as HwBridge.

These stereotypes enable us to assign non-functional properties to these hardware components. For instance, the communication bus provides a bandwidth of 25 kb/s.

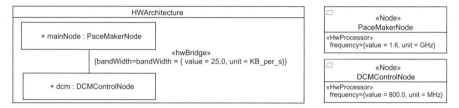

Figure 9.3. *Description of the execution platform of a pacemaker*

9.2.3. *Description of the deployment model*

The deployment of an application consists of defining the instances of the components, their configuration as well as their allocation to an execution node.

The UML composite structure that we have used for specifying the component-based model earlier defines the roles that each component plays within the system. A role that is specified at the type level is different from an instance. The entities that will run, thus the entities that we will deploy, are instances of these components and do not represent the components in themselves. In UML, an instance of a component is defined by an "InstanceSpecification", the instances of the properties of the component are thus instantiated by the "slots" that we associate with certain values. In the case where a property is typed by a component, the value associated with the "slot" is an instance of this component. This structure of hierarchical decomposition needs an instantiation on several levels, which can, at times, be too laborious to build manually. This task can be facilitated by tools that generate a tree of instances.

A system that is based on components often needs a configuration. The configuration attributes can, for instance, define the frequency of a "Timer" component in order to describe the triggering of events connected to the respective "Timer". The values of these configuration attributes can be defined over two levels:

– The attribute declaration level by providing a default value.

– The instance level, where a value is specified by the corresponding slot.

Once the components are instantiated and configured, we can proceed to the allocation of the component instances.

Broadly speaking, the allocation phase consists of defining the relation between the instances of the software components and the hardware resources of the platform. Figure 9.4 describes the desired deployment of the components

DeviceControlMonitor and PGController on the hardware platform. The relation "Allocate" provided by MARTE is used for the description of this allocation model. The generation of the instances and the deployment of these instances will be carried out throughout the code generation phase described in section 9.4.

The allocation can also be more finely tuned by introducing platform software resources at an intermediary level, between the applied components and the hardware platform. The software execution resources are tasks modeled in MARTE via the concept "SwSchedulableResource". The allocation model will consist of allocating the instances of the application components on the software execution resources, and then to allocate these software execution resources to the hardware execution resources. The "Allocate" relationship of MARTE is used to capture these two allocation levels. A more detailed consideration of software platform is also feasible (management of shared resources, etc.). The degree of specificity of the implementation platform model (which is described in the following section) depends on the level of detail of the input platform model.

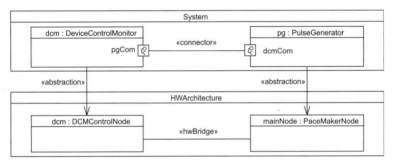

Figure 9.4. *Deployment of the software components on the platform*

9.3. Generation of the implementation model

This section describes the two main design patterns we need to apply to the component-based input model, in order to generate an component-based implementation model. The latter details how to realize the interactions between the components as well as the technical aspects such as the protection of passive components (i.e. PpUnit concept in MARTE). The proposed transformations exploit the information declared in the input model that specifies the use of interaction mechanisms (connector) and the services of a container. The objective behind these two mechanisms is the separation of concerns: the functionality of a component must be specified independently of the technical constraints of the environment where it is embedded. Section 9.3.1 introduces the main concepts that are being implemented in these transformations.

9.3.1. *Main concepts*

The concepts necessary for the generation of the implementation model are the components, the connectors and the containers.

A component is a software entity represented by a UML class. A component has the interaction points called ports. It must explicitly declare which services are provided and which services are required by the component via which port. To maximize the reuse, a component should not know the components that offer the services it requires. A component can have an explicit internal structure made up of subcomponents called assembly components.

These assembly components are represented by *parts* or UML properties. These properties are typed by components (UML classes) and have ports that can be interconnected with the ports of the component that encompasses the ports of the other subcomponents. A component can be a "type", that is an abstract class defining a set of ports, or an "implementation", which carries out the services offered by its ports.

A connector is a UML element that connects at least two *parts* or the ports of a composition. In UML, a connector denotes an interaction between two elements without providing information on the manner in which this interaction is carried out. In the way we are using connectors, this information is declared on the connector (synchronous/asynchronous communication, etc.). As a result of applying the connector pattern, as is described in the section 9.3.2, a component that carries out the interaction will be generated in the component-based implementation model.

A container is an entity whose role is to encapsulate a component (external membrane) by managing the non-functional aspects that are necessary for the execution of the component's services. Thus, the component itself only has to implement the business code.

9.3.2. *Connector pattern*

The connector pattern is based on the work of Shaw and Garlan [SHA 95] where the realization of an interaction is a first class citizen of the model. Just like a component, an interaction may have several possible realizations and configurations. The idea of the connectors was formalized within UML in [ROB 05a] and [ROB 05b]: a UML connector – a "line" in a composed class – is replaced by an interaction component.

The interaction component is typically defined in a model library as a *template*. The latter is necessary, because the interaction component must be adapted to its usage context. For instance, in the case of a call for an operation offered by a component, the interaction component must provide this operation. Mainly, we consider the case where the parameter of the *template* is an interface or a *data-type*. If it is an interface, the provided ports and the required ports of the connector are typed with this interface. At the level of realization, the implementation must also be adapted in order to correspond to the interface. The implementation of a *template* is provided as a text *template* written in a model to text transformation language (M2T).

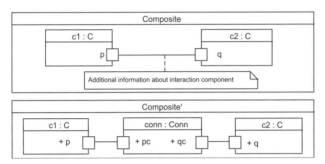

Figure 9.5. *Reification*

In our study, the majority of interactions are based on operation calls, which are without any particular support. The processing of the signals is different, because it does not have any operations that can be called. A port that allows the emission of a signal is thus associated with a *derived interface* that provides a method (called push) enabling the emission.

```
interface D_Push_SensedPulse {
  push(in data : SensedPulse);
}
```

The *template* is defined at the level of a UML package in order to enable the specification of a set of elements that depend on the formal parameter. Figure 9.6 shows the specification of two types of connectors for the data flow, one for the *push–push* model and one for the *push–pull* model. In the first case, the producer plays an active role (it produces data at a particular instant) and the consumer plays a passive role (it is called by the environment when a piece of data arrives). In the latter case, the consumer plays an active role (it reads the data at a precise instant, i.e. it *pulls for data*). The package defines two implementations for the push–pull model, one based on a queue and another one that memorizes the last value received (a queue of size 1). The generalization between type and implementation is more easily expressed with the *template* at the level of the package: there is no need for binding or inheritance.

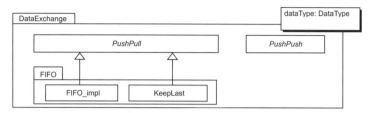

Figure 9.6. *Package template for FIFO and* KeepLast *connectors*

The use of a formal parameter is described in Figure 9.7: the ports of the abstract interaction component PushPull are typed with the formal parameter, and an attribute inside the First-In First-Out (FIFO) queue is typed with this parameter. The *template* parameter thus appears in the operations provided by the First-In First-Out (FIFO) (this is not visible in the composite structure diagram), once in the operation declaration and once in the behavior. This behavior can be described via state machines, UML activities or with a classical programming language (as an *OpaqueBehavior*). For the realization of the FIFO, the latter option has been used. A model transformation language that transforms the models to texts, such as Acceleo, may be used to instantiate the code (a text block) with the current parameter of the *template*. The implementation of this option is described in the case of state machines in the following section.

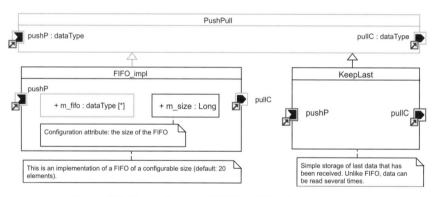

Figure 9.7. *The realization of FIFO and KeepLast connectors*

9.3.3. *Container pattern*

A known pattern for changing the way in which an object interacts with its environment is the *container*. This pattern was also identified by middlewares such as CORBA Component Model (CCM) [OMG 08] and Fractal [BRU 04].

The container encapsulates the object (the component) and can provide services and observe or manipulate the interactions of the components. The content of the container is added to the application model in a later transformation step instead of putting it directly into it. This enhances flexibility, since the content of the container changes whenever its declaration changes. This way, the final user applies a transformation rule, without knowing in detail the elements that will be added to the container.

Figure 9.8 shows this principle: component "C" is enhanced with the rules that need to be applied. This information is evaluated by a transformation of the model that creates the container and adds the elements associated with the rules. The component becomes an executor, that is the business code behind a component. We can distinguish between two types of elements in the container: the interceptors, or the extensions. The interceptor is placed on the delegation connection between a port of the container and a port of the application component that contains the code to be executed. The extension is an extra element that can be connected with the ports that are external to the container (according to its specification in the container rule).

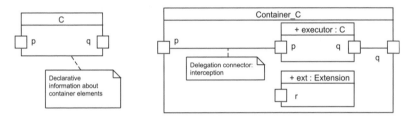

Figure 9.8. *Expansion of a container*

For instance, the container libraries offer the production of traces or the realization of the mutual exclusion pattern (just like the semaphores enable the management of the simultaneous access to a single resource). In the case of the pacemaker, one container service is particularly important: the support of state machines.

A state machine is a combination of three elements: the container, the machine itself, a *pool* of events and the interceptor that provides the events associated with operation calls (*CallEvent*). The rule that produces the state machine in the container is described in Figure 9.9. The three elements, the interceptor, the pool, the state machine as well as their connections, are modeled here. Furthermore, a stereotype of the interceptor defines the set of intercepted ports: all ports, a specified set, all input ports or all output ports. In the case of a state machine, all ports are intercepted and the interceptors produce events that correspond to the calls sent to the executor ("CallEvents" according to the UML standard).

Somewhat similarly to the FIFO connectors, the state machines and the interceptors are defined in *templates* at package level (Figure 9.10). The model of the state machine is defined in the component (the class), namely in this case study the class "Cardiologist". The instantiation of the implementation that can execute a state machine thus depends on a formal parameter of type class. The message interceptor is typed with an interface because it is placed between the typed ports with a specific interface.

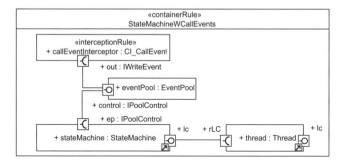

Figure 9.9. *Container rule for a state machine*

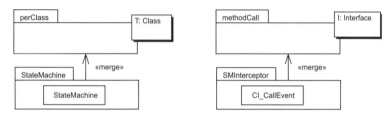

Figure 9.10. Template *of a state machine*

Figure 9.11 describes the operation pattern that (among other things) generates the execution of a state machine. The code, hereby specified in C++ language, is modeled by an opaque behavior in UML. Access to the model elements and access to control elements are embedded between [and /], for instance [name/] allows access to the name of a UML element. This code sample shows the capability to adapt the behavior of the components embedded in the container in order to create new functionalities that are completely separated from the operational code.

Figure 9.12 illustrates the result of the instantiation with the "Cardiologist" class of the case study. This class contains a state machine with the state "StartingTherapy", which appears in the "switch". In each state, the events defined as (*triggers*) for the transitions are compared with the event received by the *pool* (which is fueled by the interceptors or by a *timer*), and a new state is assigned to the current one.

9.3.4. *Implementation of the components*

A component implementation can be provided by operations whose body can be modeled with an activity described by Action Language for Foundational UML (ALF) action language or with an opaque behavior described in C++ programming language. In the case of the action language, there are specific commands for interacting with an existing port.

```
// checkPreCond body - Generated by eC3M
PR ("IN [clazz.name/]::checkPreCond(): currentState : ",
            << [clazz.name/]_CsStr(m_currentState) << showI);
[for (sm : StateMachine | ownedBehavior->select(oclIsKindOf(StateMachine)))]
int newState;

switch(m_currentState)
{
    [for (state : State | sm.region.subvertex->select(oclIsKindOf(State)))]
    case [clazz.name/]_[state.name/]:
    ...
    [/for] ...
```

Figure 9.11. *Template for the realization of operation "accept"*

In the case of third-generation programming languages such as C++, the notion of a port does not exist. In the case of an emission, an implementation can then call an operation associated with a port, or in the case of a receipt (in *push* mode), one of the operations can be called. A port thus has an association between its type and a provided and required interface. For example, a MARTE flow port that produces data is "mapped" on a "push" operation, with the transferred piece of data within parameters (type of port).

A *flow port* that consumes the data has two different realizations. In the first realization, the consumption is triggered by the environment, which calls for an operation from the consumer (executed by a task of the middleware or the caller). In the second realization, the active component verifies at a given instant if a piece of data has arrived at its port and executes it. In this case, the producer and the consumer have ports with a required interface. To couple them, we need a FIFO. In our case study, the second version is used predominantly.

```
PR ("IN Cardiologist::checkPreCond(): currentState : ",
            << Cardiologist_CsStr(m_currentState) << showI);
int newState;

switch(m_currentState)
{
    case Cardiologist_StartingTherapy:
    ...
```

Figure 9.12. *Result of the instantiation of the template with the class "Cardiologist"*

9.3.5. *Resulting implementation components*

The result of the execution of these two transformations is an implementation model that is still a component-based model. This model explains the solutions chosen for the connectors and comprises containers for all its components. This model depends on the platform and describes the realization of the MARTE input model. It is the result of the application of the component realization patterns, the container patterns and the connectors.

Figure 9.13 illustrates the implementation model that results from the class "Cardiologist" of the case study. The generated container has four interceptors, corresponding to four class ports. Each interceptor has a different type, that corresponds to different instantiations of the *template* at the package level. Moreover, the container has a *pool* of events and an instantiated state machine.

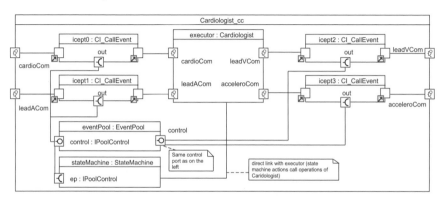

Figure 9.13. *Generated container for the "Cardiologist" class*

9.4. Code generation

The result of the previous phase is a component-based model of the application, to which we have added reified connectors and expanded containers. The code generation starting from this model needs the following two actions to occur:

– The deployment of the components. This implementation consists of breaking down the global model in submodels for each execution node. The interdependencies that the submodels have between each other must be generated. The result of this activity is, according to the deployment declared at the input, an implementation model for each execution node.

– The transformation of the notions of ports and connectors that do not have an equivalent concept in the object-oriented programming language (such as C++).

The following subsections exemplify these two activities in more detail using the pacemaker case study.

9.4.1. *Deployment of the components*

In this activity, we must produce one model per node. To do that we must separate the instances and copy the elements needed to achieve these instances.

The first task may seem insignificant in those cases where each component instance is explicitly associated with one single node. However, more complex cases may crop up. This is the case for composites components, where the subcomponents are deployed on different nodes, the mother-component is implicitly allocated to these different nodes. This imposes some constraints on the properties of these components. To avoid any issues with managing the coherence of the values of these properties, just a read-only configuration of the properties will be authorized. These constraints for well-formedness are verified in the validation phase of the input models.

The second task consists of complementing each of these models with the elements that they depend on for their execution. These elements can be types/attributes implantation, inherited classes, etc. The result of this task for each node is an autonomous model in the sense that it contains all of the necessary elements for its execution.

Figure 9.14 represents the models for the execution nodes presented in Figure 9.4. Two application models are being generated: a model for the node on which the component PGController is being deployed (and all of its subcomponents), and a model for the node where the DeviceControlMonitor is being deployed. Each of these models contains a part of the global component System and the types used by its components (DomainTypes).

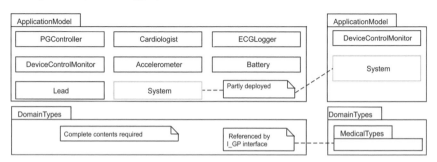

Figure 9.14. *Carrying out the deployment*

These executable component-based models can now be translated into object-oriented models for code generation in an object-oriented programming language.

9.4.2. *Transformation into an object-oriented model*

The notions of ports and connectors do not have equivalent concepts in object-oriented languages. It is thus necessary to map these concepts to a set of programming language constructs defined in a pattern. The main concepts of object-oriented programming languages are the classes, the interfaces, the attributes and the operations.

We can distinguish between the ports with provided interface and ports with a required interface (there are also ports that offer and require interfaces at the same time). A port that provides an interface is an access point to a service. To execute a component that provides this service, it is important to be able to obtain a reference to that port. If a component has a "p" port that provides an interface "I", the execution of a component must have an access operation to port "get_p", which returns the reference for that port. The implantation of this operation can be calculated automatically: if there is a delegation connector toward a *part* in the interior of the component, the reference to this part is returned; if not, the reference of the component itself is returned.

A port with a required interface is an interaction point that needs a reference toward another component that provides the interface (more precisely to the port of the component that provides the interface). The component must thus store this reference and provide a "connect" operation that initializes it from a passed value. This operation is typically called "at system startup", when connections between components are established. Thus, each port "q" with a required interface is transformed into a "q" attribute that saves the reference toward a port providing the interface and a "connect_q" operation that helps configure the component. Figure 9.16 shows the class that results from this transformation for the component "PGController", presented in Figure 9.15.

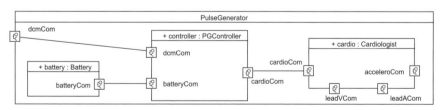

Figure 9.15. *Sample of the component-based architecture of the pulse generator*

PGController
get_dcmCom(): D_FeatureBasedCS_p_I_DCM
get_batteryCom(): D_FeatureBasedCS_p_I_Battery{nonunique}
connect_batteryCom(in: D_FeatureBasedCS_r_I_Battery)
get_cardioCon(): D_FeatureBasedCS_p_I_Cardiologist
connect_CardioCom(in: D_FeatureBasedCS_r_I_Cardiologist)

Figure 9.16. *Transformation into an object-oriented paradigm*

The code below shows the code of the operations get_dcmCom and createConnections() that is generated taking the architecture presented in Figure 9.15 as input. The first operation is associated with the delegation connector between the PulseGenerator and the controller inside – the second to the internal (assembly) connections within the pulse generator. Please note that certain connections are bidirectional, for instance the connection between battery and controller. The associated ports are typed with MARTE client/server ports that provide and require operations. There are therefore two pairs of provided–required interfaces and two connections.

```
void PulseGenerator::get_dcmCom() {
  // realization of delegation connector
  return controller.get_dcmCom();
}

void PulseGenerator::createConnections() {
  // realization of connector <controller-cardio>
  controller.connect_cardioCom (cardio.get_cardioCom());

  // realization of connector <battery-controller>
  controller.connect_batteryCom (battery.get_batteryCom());
  battery.connect_batteryCom (controller.get_batteryCom());
  ...
}
```

9.4.3. *Generating code*

Given that the transformation from component-based to object-oriented models has been already carried out, a classic code generator that takes an object-oriented UML model as input is sufficient for the generation of object-oriented programming language (i.e. C++ in our case).

For each class or interface, a C++ class is generated. The UML packages are translated into "namespace" declarations in C++. The organization of the files follows the same scheme as the one that is applied in Java – a *namespace* corresponds to a directory and the file structure thus reflects the hierarchy of UML package.

The dependencies towards the external package are translated into "include" directives to header files of libraries and a suitable linker configuration. The generated C++ code can be compiled in a C++ compilation environment.

9.5. Support tools

The *embedded Component Container Connector Middleware* (eC3M) [RAD 09] is an environment that supports the process described in the previous sections. eC3M offers validation rules as well as transformations of component models (enriched with a specification of container rules to apply and interaction components to use) toward an implementation model from which executable code can be generated. eC3M is integrated in the UML Papyrus modeling tool. Figure 9.17 gives a global view of the eC3M environment. This tool is superseded by Qompass designer.

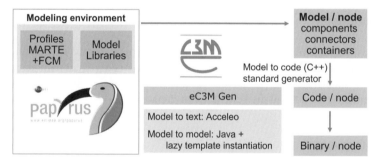

Figure 9.17. *Toolchain eC3M*

The input models are UML models that apply UML MARTE and an Flexible Component Model (FCM) profile [JAN 11]. FCM is a profile that enables us to annotate ports, connectors and components. An FCM port is characterized by a type and a kind. The kind helps to select suitable translation rules for interfaces provided/required, as we have detailed in section 9.3.4. For UML connectors, the type or implementation of the interaction component to use is described with the help of FCM. The application of the FCM connector stereotype stores a reference toward an interaction component defined within a model library that typically contains a set of interaction components for different purposes. Furthermore, FCM allows the application of a set of container rules for components. New types of interaction components and containers can be defined in libraries. Thus, the functionality offered by eC3M can be extended by extending the modeling libraries instead of the tool itself.

Starting from a UML/MARTE/FCM model, the eC3M environment carries out a set of model transformations, specifically the reification of connectors toward interaction components, the extension of the containers, a distribution of the model toward a model for each node and the transformation of the component-based model into an object-oriented model, as previously described. The code generation is based on a model generator towards a standard code, namely C++. The compilation also uses standard compilation tools, in our case the g++ gnu compilation tool. The integration in the Eclipse environment is done by using CDT, the C++ development tools for Eclipse. eC3M generates a CDT project per node and configures the paths (-I and -L) for finding the required external libraries. This chain thus enables the generation of one binary per node. eC3M currently supports a static deployment, that is a deployment without dynamic allocation. The actual deployment requires copying each binary to the node on which it is allocated. It thus suffices to copy each binary on its node in order to execute the application.

9.6. Conclusion

In this chapter, we have presented a generation process for the executable binaries starting from a MARTE model with distributed components. The main stages of this generation chain are the validation of the input models, the generation of an implementation model and the generation of the code.

9.7. Bibliography

[BRU 04] BRUNETON E., COUPAYE T., STEFANI J., The fractal component model, 2004. Available at fractal.objectweb.org/specification/.

[JAN 11] JAN M., JOUVRAY C., KORDON F., et al., "Flex-eWare: a flexible model driven solution for designing and implementing embedded distributed systems", Software: Practice and Experience, vol. 42, no. 6, 2011.

[OMG 08] OMG, CORBA Component Model Specification, version 4 (part of the CORBA 3.1 Specification), 2008. Available at OMG Document formal/2008-01-08.

[RAD 09] RADERMACHER A., CUCCURU A., GERARD S. et al., "Generating execution infrastructures for component-oriented specifications with a model driven toolchain – a case study for MARTE's GCM and real-time annotation", 8th International Conference on Generative Programming and Component Engineering (GPCE'09), ACM Press, pp. 127–136, 2009.

[ROB 05a] ROBERT S., RADERMACHER A., SEIGNOLE V., et al., "Enhancing interaction support in the CORBA component model", in RETTBERG A., ZANELLA M.C., RAMMIG F.J. (eds), From Specification to Embedded systems Application, IFIP TC10 Working Conference: International Embedded systems Symposium (IESS), Springer, pp. 137–146, 2005.

[ROB 05b] ROBERT S., RADERMACHER A., SEIGNOLE V., *et al.*, "The CORBA connector model", *Proceedings of the 5th International Workshop on Software Engineering and Middleware*, ACM Digital Library, 2005.

[SHA 95] SHAW M., GARLAN D., *Software Architecture: Perspectives on an Emerging Discipline*, Prentice Hall, 1995.

PART 4
AADL

Chapter 10

Presentation of the AADL Concepts

10.1. Introduction

This chapter presents the second version of the *Architecture Analysis and Design Language* (AADL) architecture description language. AADL is a standard defined by the SAE International (formerly Society of Automotive Engineers) [SAE 09]. It is a part of the family of architectural languages that help structure one or more implantations of a system, in order to facilitate the analysis of the system via different techniques (simulation, formal method, model-checking) including its realization.

In section 10.2 we will introduce the general concepts of architectural description languages (ADLs), and then in section 10.3 we will give a high-level presentation of the AADL version 2 (AADLv2) and the associated tools. We will then detail the language in section 10.4 and its appendices in section 10.5. Finally, we will conclude with a brief overview of the analyses that are supported by AADL in section 10.6.

10.2. General ADL concepts

Several ADLs have been proposed, each of them providing different levels of abstraction, depending on the documentation needed, the analysis requirements or the requirements for code generation. For this reason, the precise definition of an ADL has been rather vague for a long time, unlike classical programming languages. The work of Medvidovic and Taylor [MED 00] provides the first definition of what an ADL is, a classification which can be summarized as follows: "An ADL enables the description of the architecture of a system as a set of components interlinked via

Chapter written by Jérôme HUGUES and Xavier RENAULT.

connectors that describe the interaction mechanisms, thus defining a configuration of the system".

A component is a composition unit whose interface and execution context are wholly analyzed, for instance an Ada task and a Java EJB component. The connectors define the semantics of the interaction between the interfaces of the two components, for instance a procedure appeal and a message sending. The configuration of the system is a set of components and connectors that are syntactically and semantically coherent.

Having presented these definitions, we may now proceed to introducing the basic concepts of the AADL.

10.3. AADLv2, an ADL for design and analysis

The AADL is standardized by the SAE [SAE 09]. AADL allows us to model both the software and the hardware aspects simultaneously, for a real-time system, by supporting a modeling procedure done via refining (simple heritage), with a clearly defined semantic that is in accordance with the avionic and spatial systems, and an advanced mechanism for managing the non-functional properties.

10.3.1. *A history of the AADL*

AADL has a long history, inherited from several successive projects that have sought to clarify the perimeter of an ADL. This long maturing process guarantees not only a higher expressive ability, but also guarantees that the model built can be analyzed automatically with a computerized tool. AADL is born out of two "historical" ADLs, whence it borrowed several concepts: MetaH and ACME.

MetaH [FEI 00] is an ADL initiated by Honeywell Technology at the initiative of the Defense Advanced Research Projects Administration (DARPA). Its main author, Steve Vestal, drew inspiration from the syntax of the Ada language for its textual representation. MetaH was espoused of explicit concepts for modeling tasks, subprograms, processors, etc., as well as assigning non-functional properties to these concepts. MetaH was used in several American military projects, especially avionic projects. At the same time, MetaH was supported by a modeling environment, tools for analysis and code generation toward hybrid automatons, besides also covering the activities of the software's lifecycle.

ACME [SCH 04] is an ADL developed at the *Software Engineering Institute* (SEI) of Carnegie Mellon University, disposing of extension capacities and the ability to refine its components, as well as of having mechanisms that are specific to real-time systems, therefore remaining quite generic.

The explicit manipulation of concepts close to the engineer provided by MetaH and the modeling capacities obtained by the extension and refinements mechanisms of ACME are the two building blocks of the AADL. They facilitate both the modeling of large-scale systems, benefiting from its ability to extend close to the object modeling, but also the modeling of precise systems, where all of the elements that are relevant for the analysis or for the code generation are present in the model.

The efforts of standardizing the AADL language started in SAE in 1999, in order to make available an open standard, and thus facilitating the construction of integrated sequences tool chains. The goal of the AADL language is to allow the simultaneous designing of the architecture, the analysis (verification and validation) and generation of systems, all in a real-time context. The objective is to have a textual syntax, a graphic representation and an XML-based exchange format, deriving from a metamodel.

The first version of the standard was published in 2004, following a joint effort of the SEI, *Department of Defense* (DoD), Lockheed Martin, Airbus and the European Space Agency (ESA). The AADL committee soon expanded, joined by the academic community that built bridges between formal analysis tools and AADL.

A second version of the standard was published in 2009, expanding on the one hand the expressiveness of the language, allowing for more modularity (reinforced use of packages, parametric modeling, abstract components, etc); and on the other hand, new annexes were defined. They provide guidelines for the advanced modeling of the data, for the behavior of the components or of the errors, for avionic systems based on the standard ARINC653 [ARI 07]. It is this second version that we will be tackling here.

It is important to emphasize that the development of the AADL was made in connection with several research and development projects, combining academic and industrial users alike, among which IST-ASSERT, TopCased, SPICES, the SAVI initiative of the *Aerospace Vehicle Systems Institute* (*AVSI*)[1]. The evolutions of the language took into account the different needs expressed by these projects.

10.3.2. *A brief introduction to AADL*

In this section, we present the fundamental elements of an AADL model, which will be detailed at a later stage.

Figure 10.1 provides a high-level view of the different blocks that form an AADL model. They are made of property sets and packages. Property sets define the

---

1 See `https://wiki.sei.cmu.edu/aadl/index.php/The_Story_of_AADL` for more details on this project.

attributes that can be attached to several elements of the mode: name, type and unit. The packages bring together the elements of the model: component types and component implementation.

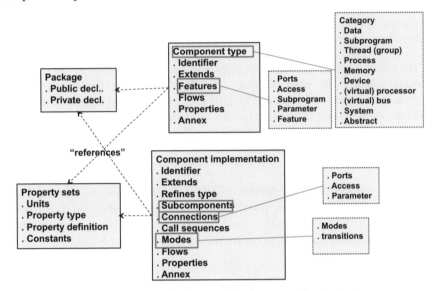

Figure 10.1. *Simplified view of the elements of the AADLv2*

– The component types define the signatures of the components forming the basic bricks of the architecture: its identifier, the component it extends through a simple heritage mechanism, its interface and its properties. The types correspond to a predefined category (tasks, processors, etc.); the interfaces with known interaction mechanisms: message exchange, event exchange, data exchange, etc.

– The component implementation defines the concrete relations between the types, its subcomponents and the connections between these subcomponents and the interface of the components. An implantation has its own property set.

Within a component, the modes allow us to model the different configurations of a system and the reconfiguration decisions in case of an unforeseen event (for instance, in case of an error, or of a boot command). Each configuration can specify different values for the properties, but also activate/deactivate certain components. Thus, the modes define the static set of configurations of the system. AADLv2 is therefore an ADL in the framework of the [MED 00] taxonomy.

The different elements of a model must be grouped in packages, defining as many spaces as there are different denominations. The dependencies between packages must be made explicit throughout the modeling process, as is usually the case with programing languages.

AADL defines a set of syntactic and semantic rules that guarantee the coherence of the models and their well-formed composition. These rules guarantee that the model is made up of a valid set of components, but do not help it suppress the need for verification and validation activities supported by tools or by a review of the model.

10.3.3. *Tools*

A modeling language without a tool has a very limited scope. To accurately test the language, AADL was implanted in different tools, some of which are pieces of open source software:

– OSATE2: the SEI develops a reference implantation of an AADL tool, OSATE [CMU 09], as an open source Eclipse extension. OSATE negates the requirement for a front end that supports AADLv2 language as well as an advanced text editor. By benefiting from the Eclipse technology, OSATE facilitated the development of numerous gateways between AADL and the analysis tools: model optimization tool, security analysis tool or flow latency calculation tool;

– AADLInspector: this is a piece of software developed by Ellidiss Technologies. It provides a small text editor for editing AADLv2 textual models, as well as connections toward different analysis and simulation tools. It is based on Cheddar [SIN 07] on the analysis of the scheduling and a tool that facilitates the simulation of complete AADLv2 models through a behavioral description;

– Ocarina: Ocarina [LAS 09] is a tool suite designed in Ada by the TELECOM ParisTech S3 team, with the help of the Higher Institute of Space and Aeronautics (ISAE) as well as the ESA. Ocarina was designed as a traditional compiler divided into two parts: a front end and a family of back ends for generating code and for analyzing models. Thus, Ocarina provides a code generator for the PolyORB-HI C and Ada middleware, the colored Petri nets or the timed networks, and annotations for the worst-case execution time analysis tool (WCET) *Bound-T* task models for the scheduling analysis with Cheddar and a language of constraints as annex language, REAL, allowing us to restrict the modeling patterns used.

10.4. Taxonomy of the AADL entities

In this section, we will present the different AADL language elements in these graphic and textual notations.

Revisiting the canonical elements of an ADL, AADL defines the notion of components and connections. The connections are a simplified version of the connectors for which only certain policies are supported. Components and connectors alike are connected in the implantation components thus forming a configuration. Components, connections and configurations can be annotated with properties that refine their description.

In the AADL terminology, a property is an element which, through its value, characterizes an AADL component: task priority, scheduling policy, memory size, task triggering policy, etc. It thus differs from the notion of system property, which is obtained by analyzing a submodel of the model, submodels such as the fact that it can be scheduled, secure, etc.

10.4.1. *Language elements: the components*

AADL distinguishes between two families of components: the software components and the hardware components.

The software components allow us to describe the applied elements of an architecture. The hardware components describe the elements where the applicable elements are used. These components are organized in a hierarchical fashion in hybrid components called "systems", which can themselves contain other subcomponents, or can contain "abstracts", when the category of the component is not yet known.

10.4.1.1. *Software components*

We may distinguish several types of software components:

– the tasks (*threads*): the tasks model a competing task or an active object. There is a scheduling element that can run at the same time as the other tasks attached to the scheduler. It always runs in the memory space and it is always triggered either periodically on a temporal event, or upon the arrival of data or events on their interfaces (*ports*). The latter are frozen once the execution has been triggered: no event or data can be considered until the next trigger;

– the task groups: they enable the logical grouping of the tasks of a system (*pool* of tasks, for example). A group of tasks is contained in a system component and may contain, besides the tasks themselves, data components (variables shared by the tasks), or even components that represent subprograms (called by the tasks of the group, when they are activated);

– the processes that represent an addressable memory space; they may contain data components, subprograms or tasks;

– the subprograms: a component of this type represents an execution sequence called with parameters and has no proper state (no static data, for example). The subprograms, just like tasks, can be organized in groups thus representing libraries;

– the data representing static data. Several components can share the same resource, their mutual exclusion thus being part of the exigencies.

The graphical representation of each of these components is presented in Figure 10.2.

Figure 10.2. *Graphic syntax of the AADL software components*

10.4.1.2. *Hardware components*

The hardware components of an AADL model allow us to specify the elements where the application will be used. We can distinguish between:

– the processors, which are the hardware units that enable the execution of the tasks. They may contain and have access to the memory, communicate with peripheral devices or with other processors via the buses;

– the memories represent elements that allow us to record the code or binary data. We may also speak of a RAM component as well as of a more complex hardware component, such as a card. Their typical attributes are the number and the size of their addressable space;

– the buses are elements that allow the exchange of control flows or data flows between the processors, the memories and the peripheral devices. Typically, these are communication channels with protocols attached. They can connect directly with each other in order to create complex networks;

– the peripheral devices (*devices*) representing specific elements of the hardware, elements that are external to the system or even elements that interact with the exterior from the system outwards. These external devices can have their own processor, their own memory and software components, and they can all be analyzed separately.

To these purely hardware elements we can add the "virtual components", processors and bus. They represent an abstract view of these components: a virtual processor that represents a sub-set of a processor (partition or core of a processor, a virtual bus representing a protocol above a physical bus). The graphic syntax of each of these components is presented in Figure 10.3.

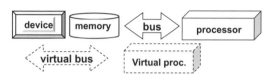

Figure 10.3. *Graphic syntax of AADL hardware components*

10.4.1.3. *Hierarchical characterization via the systems*

Throughout the modeling process, several allocation decisions may remain pending. Therefore, we may know the interface of a component without, however, knowing its category. The "abstract" components allow us to represent these components. These components cannot be used as such in the final configuration of the system: they must be refined in a more concrete category.

The hybrid components are organized according to a hierarchy that represents the global structure of the model. This involves a set of software components, hardware components and other systems that interact with each other. The graphic syntax of a hybrid component is presented in Figure 10.4.

Figure 10.4. *Graphic syntax of hybrid components of an AADL model*

10.4.2. *Connections between the components*

We have presented the different types of components found in an AADL model. They interact due to their interface elements. The features of a component specify the manner in which it interacts with the rest of the system via:

– ports, which are interfaces allowing components to exchange events or data. We classify them in three different categories:

- *data ports*,

- *event ports*,

- *event data ports*.

The ports only have one direction: an exit port will be connected to an entry port. We may distinguish between "*in* ports", "*out* ports" and bidirectional ports (*inout* ports);

– access to subprograms, modeling the calls from external components of an instance of a subprogram, for instance a rendez-vous between tasks or a distant procedure call;

– parameters, which are data values passed upon the entry (or the exit) of the calls to the relevant subprograms;

– access to these data, specifying the access to a shared variable;

– access to buses that represent the physical connectivity of processing components, memory and other components.

Certain characteristics of a component can be required (*requires*) or provided (*provides*), enabling the definition of contracts between components.

Throughout the definition of an implantation of a component, these subcomponents are connected between them and at the interface of the mother component. This connective weaving thus allows us to build a configuration of the system.

Figure 10.5 describes the graphic syntax of these different interface elements.

Figure 10.5. *Graphic syntax of the interface elements of an AADL analysis*

10.4.3. *Language elements: attributes*

The AADL attributes, as they are presented in the second version of the standard [SAE 09], can be divided into several categories:

– the deployment attributes: components connected to a processor, a memory;

– the attributes connected to communication elements: the size of the message queues, thread managing policy, message priority, processing of incoming messages, message dissemination policy;

– the attributes connected to stocking elements: memory size, heap size;

– the programming attributes: procedure calls, priority;

– the task-related attributes: priority, WCET;

– the temporal attributes: deadlines, communication delays.

The possible combinations of AADL attributes are very rich and have been defined in order to permit several analyses. We hereby present a somewhat incomplete classification of these attributes, and we connect them to the corresponding possible analyses.

10.4.3.1. *Deploying attributes*

These attributes allow us to analyze the linking constraints that may exist between a software component and a hardware component (or a hardware category).

For every attribute considered, the standard defines three variations around one key word, which we will note χ:

– `Allowed_`χ`_Binding_Class`: allows us to analyze the classes of elements that can be linked to the software component as a set;

– `Allowed_`χ`_Binding`: set of hardware components that can be linked to the software component;

– `Actual_`χ`_Binding`: defines the hardware material that is connected to the software component for a configuration.

The key words that have been considered in the standard are: `Processor, Memory: Connection Subprogram_Call`.

Besides these attributes, we also find:

– an indication that several components are used and connected on the same material (software regrouping of the components);

– we are able to specify the service quality expected by the connections;

– an indication that the policies are put into practice (trigger for the processors, read-only mode for the memory, task scheduling);

– an expression of the limits, such as the maximum number of tasks managed by a processor and the priority terminals that we can attach to them.

The attributes relative to the use can be exploited in different ways for the analysis of the system:

– validation of the adequacy of the components used in relation to the sub-hardware (observing the expressed constraints);

– validation of the security of the system: communications flow within the system;

– information for building the architecture of a formal model (interactions between software components and resource components, shared or not shared).

10.4.3.2. *Attributes linked to communication*

These attributes allow us to characterize a communication that takes place between different components. We can distinguish between:

– service quality policies:

- connected to sending messages (`Fan_Out_Policy`): when a message is sent, it can be broadcast to everyone else, it can choose a recipient (rotationally, that is round robin), or it can send the message on demand;

- connected to the arrival of new messages in the queue: if the queue is full, Overflow_Handling_Protocol allows us to delete the newest or the oldest message in the queue, or to eliminate an error;

- connected to the parametering of the connection: the size of the queue (Queue_Size), or even the type of transmission (*push, pull*).

– information regarding delays, sending rate or receiving rate:

- (Output_Rate, Input_Rate): number of messages sent or received per second (or per task trigger); (Output_Time, Input_Time): the moment where the messages are sent or received (before the task calculation, after, during or at the due date);

- transmission (Transmission_Time);

– attributes characterizing the latency of the system (from one end of the system to the other, for the data flow).

These attributes offer the possibility to analyze the communications of the system: the safety of the communications in terms of message loss, delivery time, observing the due dates.

10.4.3.3. *Attributes relative to the storage components or the transfer components*

We can define the elements in order to size the tasks: the size of the heap (Source_Heap_Size), the size of the stack (Source_Stack_Size) and the size of the data (Source_Data_Size).

The tasks are connected (via the hosting process) to a physical memory zone, for which we are able to specify:

– the access permissions for reading and/or writing in the memory or on the bus (Acces_Right): read-only or write only, both reading and writing or even access uniquely via calls for specific methods;

– the physical access time to the memory in writing (Write_Time);

– addresses that define the memory space.

These attributes allow us to carry out analyses on the security of the interactions between the components and on the good sizing of memory areas. These verifications are carried out using the tool Ocarina.

10.4.3.4. *Programming attributes*

These attributes offer the possibility of analyzing the behavior of the tasks upon receipt of external events, by triggering their execution, or even by connecting the source code and the AADL subprograms.

We call an *entrypoint* the reference toward a subprogram. This could be an AADL subprogram, a sequence of subprograms or of a chain of characteristics that reference a function written in a different language. These entry points are attached to particular elements of the lifecycle, as a property X_Entrypoint where X corresponds to:

– Activate: everything that regards the task when it has been selected for an execution mode;

– Compute: regards the execution of a task once it has been triggered. If we are talking about the attribute of a port, then this can be dissociated from the attribute similar to the task that has the port (triggering upon receipt of an event, for example);

– Deactivate: programs that are likely to be executed when the task has been deselected from a running mode;

– Finalize: a procedure executed by a task when the task has ended;

– Initialize: set of procedures designed to run when a task starts within the system;

– Recover: procedure carried out by the task in the case where the latter goes into error management mode (*recovery*).

These properties complete the configuration of the system by connecting the behavior to the architecture and the implantation to the subprograms.

10.4.3.5. *Attributes connected to the task*

These attributes allow us to specify the information relative to the active components of the analysis: they facilitate the research on the task triggering policies, the simultaneity and the passage from one AADL mode to another. These are classified as follows:

– the information connected to the triggering of tasks: Dispatch_Protocol (aperiodic, sporadic, periodic, etc.) and Dispatch_Trigger (specifies the list of entry ports likely to trigger the task upon receipt of a message);

– the scheduling information: POSIX_Scheduling_Policy: indicates if the system is scheduled with a FIFO policy, a RR (Round Robin – taking turns) or some other policy. The attributes Piority, Criticality, Urgency and Time_Slot are complementary to the attributes mentioned previously. An extra attribute allows us to indicate which policy we need to apply in case of mutual exclusion (`Concurrency_Control_Protocol`);

– the information connected to the processing policies of the messages received: `Dequeue_Protocol` (process one, several or all of the messages).

This information allows us to analyze the behavior of the application, the limits on the size of message threads or even the scheduling of the system.

10.4.3.6. *Time attributes*

The last category of attributes we need to present is the category of those attributes that specify the execution time connected to components such as tasks, peripheral devices or the subexecutive.

We can also specify the *deadline* and the execution time (a time interval that defines the limits of this time) for the different states of a task or of a subprogram: activation, initialization, reconfiguration, etc.

Other attributes complete the time description of the system:

– those regarding the loading time of a task in the system, the duration of a context change, latency on the communications, etc.;

– `Period`, which gives the minimal delay between two triggers of a task;

– those that regard the notion of time within the execution platform: the `Jitter` of the material clock, the time the executive has to carry out a change of context for the processes, the time it takes to change the context for the tasks of the same process, etc.

These attributes allow us to carry out several time analyses throughout the whole system, even to refine the behavioral analyses by eliminating, with the help of the time constraints they express.

10.4.4. *Language elements: extensions and refinements*

AADL supports the incremental modeling of the systems by successive refinements and extensions. The key word "extends" indicates that a component (type or implantation) extends another component. In this case, the properties, interfaces and modes of the mother component are inherited. It is thus possible to complete the interface of the component and to overload certain properties.

In certain cases, it is also necessary to overload an already existing element. AADL defines the key word "refined to" that serves to indicate that an element (interface or subcomponent) is refined toward a more concrete type.

10.5. AADL annexes

The previous section presented the core of the AADLv2. However, this language is designed to be extensible in order to enrich the complementary concepts that facilitate new types of analysis. The standard thus authorizes the definition of the "annexes". Two types of annexes are hereby defined as:

– the annex documents define the modeling guides, the sets of complementary properties, among which we can cite the annex of data modeling, or the annex ARINC653 for modeling the avionic systems according to the design patterns of the integrated modular avionic (IMA);

– the annex languages allow us to add a complementary description in a different language to the AADL model, in order to complete the description. This can be a behavioral description or a description of the perpetuation of errors within the system.

The set of these annexes are published by the standardization committee [SAE 11]. A tool is free to create specific annexes. The standard indicates that an annex cannot conflict with the standard and yet be ignored. This limitation helps us restrain from distorting the semantic of the standard.

The modeling annex data that will be useful to us for modeling our case study are presented below.

10.5.1. *Data modeling annex*

AADL enables us to define types of components whose category corresponds to a piece of data. The standard properties of the language provide several elements regarding the use of the memory or the competing access protocol to these data. On the other hand, nothing is indicated regarding the representation of these data: whether it is a simple type or a composed type, an implicit unit, a representation of the memory, etc.

The goal of the data modeling annex, published in [SAE 11] answers these questions by defining a set of properties and composition rules allowing us to precisely define the nature of the data used by a system, whether it is within the parameters of a subprogram or the ports of the components.

The `Data_Model` set defines 15 additional properties; in the following, we will present a few of these properties:

– `Data_Representation` indicates the type of basis for a piece of data (entire, table, boolean, etc.). This property authorizes the use of other properties, according to the representation retained;

– `Data_Digits` and `Data_Scale` allow us to indicate the precision of a number with a fixed comma, whereas `Number_Representation` or `IEEE754_Representation` allows us to indicate the resolution of a number;

– `Initial_Value` allows us to indicate the initial value of a piece of data;

– `Integer_Range` (`Real_Range` respectively) allows us to indicate the acceptable value interval for a given type;

– `Measurement_Unit` allows us to indicate the unit associated to a value.

These additional properties allow us to refine the compatibility analysis of the interfaces, for instance for the exchange and processing of physical measurings or for guaranteeing the precision of the calculations in a chain of receivers/calculators. The following example shows us how to model a variable that contains the value of the impedance of a pacemaker lead, expressed in ohms.

```
1   package Pacemaker_Data
    public
        with Data_Model;

        data Impedance_Lead  -- Impedance of a lead
6       properties
          Data_Model::Data_Representation => Integer;
          Data_Model::Initial_Value => ''85'';
          Data_Model::Measurement_Unit => ''ohm'';
        end Impedance_Lead;
11  end Pacemaker_Data;
```

10.6. Analysis of AADL models

AADL is an analysis language that is particularly suitable for varied analyses of real-time systems.

In classical model-based engineering, the engineer can analyze, very quickly and very easily, the different views of its system by separating its preoccupations. By behaving this way, when the validation and verification phases arise, the engineer must recover the disseminated information from all the different angles of its system. They are not necessarily orthogonal and can be useful for carrying out several types of analyses: very often, it cannot be satisfied with a particular view of its system, but needs a composition of angles or a subset of angles.

Therefore, we may ask the following question: once the useful model is produced, how and for what types of analyses should we use it for? With what tools? We have started to answer this question in the previous section. To complete these results, we will explain how the engineering of such systems is based on both the actual models and the analyses that we wish to carry out, which is also one of the main goals of the AADL.

We have presented the different categories of AADL attributes with the aim of classifying them in order to facilitate the formal analysis. We will indicate for each of these sets of attributes if they contain information that enables us to characterize:

– structural properties;

– qualitative properties;

– quantitative properties.

This is summed up in Table 10.1.

AADL attributes Impact	Structural	Qualitative	Quantitative
Deployment	×		×
Communication	×	×	×
Programming		×	×
Tasks	×	×	×
Time			×

Table 10.1. *Impact of the AADL attributes on the analysis properties*

10.6.1. *Structural properties*

They are connected to the structure of the system, such as:

– the connections and the coherence between the interfaces of the system's components;

– the invariants that need to be maintained within the system;

– the fault analyses *Fault-tree analysis* (interdependence between the components when one of them breaks down), latency analyses from beginning to the end of the information flows, energetic resources, etc.

The majority of these properties must be established very early in the development process, often times with a low granularity. They can be refined or enriched when the conception of the system evolves.

10.6.2. *Qualitative properties*

They are relative to the behavior of the system, that is regarding its scheduling, detecting famine, interlock or regarding the causality links between the components.

To process such properties, the behavior of the system must be defined. They are traditionally described later in the developing process, when the information relative to the behavior of the components becomes accessible due to the programming attributes or the behavioral appendix of the language.

10.6.3. *Quantitative properties*

These are used in order to evaluate the performances of the system or for evaluating its behavior according to probabilistic criteria or execution time. To establish this type of property, information on the execution time is required.

AADL is a description language for standardized architectures that has proven to be useful for carrying out several analyses on real-time systems: numerous works that have exploited AADL in this sense have been carried out around BIP [CHK 09], by TLA+ [ROL 09], UPPAAL [PON 07], LOTOS [HAM 07], Lustre [JAH 07], Cheddar [SIN 07] or by Archeopteryx [ALE 09]. These previous works allow us to draw a first classification of the manner in which AADL is used throughout the different phases of analyses of a system.

Table 10.2 presents for each of the elements of an AADL model a recap of the information or facilities that they bring according to the type of analysis considered. The columns categorize different types of analyses that are possible to carry out on a system. The lines are divided in two categories:

– the AADL attributes: for each AADL element, the table indicates on which type of analysis the value of the attribute will have a significant impact;

– the methods of analysis: the table indicates which type of analyses are likely to be carried out with each of them.

We can now answer the following questions:

– what type of validation or verification do we wish to carry out?

– how can we carry it out?

– what information does this analysis require?

We can also navigate in this table in different manners: what methods of analysis would this or that attribute be useful for? What can we verify or validate if we wish to use a particular technique, and if we wish to focus on certain attributes?

Table 10.2 is an essential step in a formal verification and validation process around an architecture description language such as AADL.

	Coherence of the interfaces	System invariants	Fault tree	Scheduling	Vitality	Causality Interlocks	Performance analysis
AADL elements							
Deployment	×			×		×	×
Time		×		×			×
Communication	×					×	×
Programming	×	×	×			×	
Tasks		×	×	×	×	×	×
Methods of analysis							
Simulation				×			
Semantic analysis	×			×			×
Type verification	×						
Proof of the theorem	×	×	×		×	×	
Model-Checking	×	×		×	×	×	
Temporal							×
Stochastic							×

Table 10.2. *Crossing of the attributes and their impact on the analysis*

10.7. Conclusion

In this chapter, we have presented the fundamental concepts of the AADLv2. This architecture description language was defined in order to enable a precise description of a real-time embedded system.

AADL has a double textual and graphic representation that the designer can manipulate, and an XML representation to help design the analysis tools. The language was designed on the basis of fundamental concepts for embedded systems (such as task notion, subprogram and peripherals), while at the same time offering modularity mechanisms, encapsulation and a simple inheritance form. By their definition, these concepts can be easily used as analysis tools; for this reason, several initiatives have been proposed, initiatives that help to carry out scheduling analysis, security analyses, memory resources, simulation or even code generation.

We will present several of these initiatives in the following chapters, and showcase them using our pacemaker case study.

10.8. Bibliography

[ALE 09] ALETI A., BJORNANDER S., GRUNSKE L., *et al.*, *ArcheOpterix: An Extendable Tool for Architecture Optimization of AADL Models*, IEEE Computer Society, Los Alamitos, CA, pp. 61–71, 2009.

[ARI 07] ARINC 653, Report, Aeronautical Radio Incorporated, 2007.

[CHK 09] CHKOURI M.Y., ROBERT A., BOZGA M., *et al.*, *Translating AADL into BIP – Application to the Verification of Real-Time Systems*, pp. 5–19, Springer-Verlag, 2009.

[CMU 09] CMU/SEI, Open source AADL tool environment (OSATEv2), Report, CMU/SEI, 2009.

[FEI 00] FEILER P.H., LEWIS B., VESTAL S., Improving predictability in embedded real-time systems, Report no. CMU/SEI-2000-SR-011, Carnegie Mellon University, December 2000. Available at la.sei.cmu.edu/publications.

[HAM 07] HAMID I., NAJM E., "Real-time connectors for deterministic data-flow", *Embedded and Real-Time Computing Systems and Applications, RTCSA'07*, pp. 173–182, 2007.

[JAH 07] JAHIER E., HALBWACHS N., RAYMOND P., *et al.*, "Virtual execution of AADL models via a translation into synchronous programs", *EMSOFT*, Salzburg, Austria, pp. 134–143, 2007.

[LAS 09] LASNIER G., ZALILA B., PAUTET L., *et al.* "OCARINA: an environment for AADL models analysis and automatic code generation for high integrity applications", *Reliable Software Technologies'09 – Ada Europe*, LNCS, Brest, France, pp. 237–250, June 2009.

[MED 00] MEDVIDOVIC N., TAYLOR R.R., "A classification and comparison framework for software architecture description languages", no. 26, no. 1, pp. 70–93, 2000.

[PON 07] PONTISSO N., CHEMOUIL D., "Vérification formelle d'un modèle AADL à l'aide de l'outil UPPAAL", *Génie Logiciel*, vol. 80, pp. 36–40, March 2007.

[ROL 09] ROLLAND J.-F., Développement et validation d'architectures dynamiques, PhD thesis, University of Toulouse, 2009.

[SAE 09] SAE, Architecture analysis and design language (AADL) AS-5506A, Aerospace Information Report, Version 2.0, The Engineering Society for Advancing Mobility Land Sea Air and Space, January 2009.

[SAE 11] SAE, SAE architecture analysis and design language (AADL) annex Volume 2, Report, Society of Automotive Engineers, 2011.

[SCH 04] SCHMERL B., GARLAN D., "AcmeStudio: supporting style-centered architecture development (research demonstration)", *Proceedings of the 26th International Conference on Software Engineering*, Edinburgh, UK, pp. 23–28 May 2004.

[SIN 07] SINGHOFF F., The Cheddar project: a free real time scheduling analyzer, 2007. Available at beru.univ-brest.fr/ singhoff/cheddar/.

Chapter 11

Case Study Modeling Using AADL

11.1. Introduction

As described in the Chapter 10, an Architecture Analysis and Design Language (AADL) model provides the software and hardware architecture of a real-time system, thus serving to design and analyze the respective system. In this chapter, we will describe the AADL model of the pacemaker, which will be used in the next chapters to illustrate its usability in terms of analysis and code generation.

Besides the actual description of the model, in this chapter we will explain the modeling choices that have guided us toward the architecture presented here. These choices are based on two types of constraints that are important in every modeling process: the characteristics of the chosen modeling language, and the constraints specific to the field in which our system will be used. The characteristics of the AADL language have been given in Chapter 10, and the main characteristics of the pacemaker have been presented in Chapter 2.

When necessary, we will review and detail certain properties that are expected from a pacemaker. Although Chapter 2 details the system specification of a pacemaker, our final design objective means that in this chapter we must detail its internal functioning.

Let us briefly present the information sources we started with in order to arrive at the architecture presented in this chapter.

– *Information sources regarding the functioning of the pacemaker*: With regard to the constraints related to the application domain, the basic starting point for our

Chapter written by Etienne BORDE.

modeling process is the system analysis of the pacemaker [BOS 07]. Starting from this specification and from a work that details the medical use of a pacemaker [BAR 10], we have produced the AADL model described in this chapter. The first document we have referenced details the requirements of the pacemaker and its system analysis, while the second document describes the functioning of the pacemaker in connection with its field of use: the medical field. Using these two documents, we were able to understand, on the one hand, the requirements involved in the design of the pacemaker and, on the other hand, the physiological phenomena to which these requirements are connected.

– Information sources regarding the modeling language. For design purposes, a modeling language seeks to define a vocabulary and a set of concepts that enable us to unequivocally represent the system's responses to requirements.

Such a language therefore imposes several constraints, so that the analyzed architecture is coherent and can be interpreted by different tools. These constraints were presented in Chapter 10, but here we will only review those that have guided us in our modeling endeavor. An AADL model must observe the standard structuring rules (composition, connection and deployment). For example, a component *system* can be composed of the subcomponents *system*, the *process*, the *processor* and/or *device*.

Given the structuring rules of an AADL model, we have worked with iterations, which have all followed the next guiding stages:

– Structural decomposition, deduced from the structural constraints imposed by the topology of the constitutive entities of the pacemaker.

– Modeling of the software architecture and of the behavior of its components, deduced from the functioning constraints described in the system analysis of the pacemaker. The medical motivations and functionalities associated with these constraints are detailed in [BAR 10].

– Modeling of the deployment of the software components on the hardware architecture, which describes the allocation of the software entities to the hardware resources.

Therefore, for each of these stages, the modeling must address the structuring constraints of the chosen modeling language, by responding as precisely as possible to the requirements of the modeled system.

This chapter describes the different modeling stages of our guidelines. For each stage, we start by reviewing the constraints of the system, then by delivering our AADL modeling of the responses to these constraints.

11.2. Review of the structure of a pacemaker

The first constraints in the structure of the pacemaker come from its decomposition in three subsystems, each playing very different roles, and each of them having to be implemented in completely different natural media:

1) The electrodes (called *Leads* in system analysis) aim to sense the natural activity of the heart and to stimulate its artificial activity. Therefore, the electrodes must be implanted straight into the myocardium of the patient.

2) The pulse generator (*Pulse Generator* in system analysis) is responsible for controlling the cardiac activity, aiming to both supervise the heart's natural activity, and to trigger the artificial activity, if need be. The pulse generator is thus connected to electrodes, but it can be implanted in an inert area of the human body in order to minimize the discomfort felt by the patient.

3) The supervision and configuration station (called *Device Controller-Monitor* in system analysis) enables the physician to ensure the well functioning of the pacemaker and to modify its functioning parameters. Thus, the supervision and configuration entity is not necessary for the nominal functioning of the pulse generator. It is not implanted in the body of the patient, but rather communicates with the pulse generator by means of a telemeter. This enables the physician to visualize the data registered by the pulse generator and to modify the configuration of the latter if necessary.

This initial decomposition of a pacemaker tells us something about the nature of the subsystems: the electrodes are simple peripherals that do not have computing resources; the pulse generator is an embedded system made up of both software and hardware components: and the supervision and configuration entity is a (non-embedded) computer system designed for the medical personnel that ensures the implantation and the supervision of the pacemaker.

Besides the nature of the subsystems, the structural decomposition informs us of the connections that must exist between the different subsystems. What remains to be determined is the nature of the information exchanged via these connections. Between the electrodes and the pulse generator, the information exchanged is nothing more than electrical signals. The nature of the information exchanged between the configuration station and the pulse generator is more varying and more complex: some information corresponds to the data sent by the pulse generator to the configuration station, while others corresponds to the configuration commands or supervision requests sent from the configuration station toward the pulse generator.

The following section describes the AADL model of this first level of decomposition.

11.3. AADL modeling of the structure of the pacemaker

In this section, we will present the AADL modeling of the first level of decomposition: the system structure. We have organized this presentation in two subsections. First, we will study the decomposition of the system into a set of interconnected subsystems and then we will represent the physical structure of the pacemaker, that is of the computing and communication resources that execute the features of the pacemaker. This second phase refines the first phase: each computing resource identified is a subcomponent of one of subsystems of the pacemaker.

11.3.1. *Decomposition of the system into several subsystems*

The structural constraints that we have presented in section 11.2, as well as the system analysis of the pacemaker [BOS 07] have prompted us to model the structure of the pacemaker in a set of three AADL components of the *system* category. These subsystems correspond to (1) electrodes, (2) the pulse generator and (3) the supervision/configuration station.

Starting from this decomposition and the description of the nature of the interactions between these two subsystems, we have added the communication interfaces between the corresponding AADL components.

We have chosen to describe the communication interfaces between the electrode subsystem and the pulse generator by means of the AADL event ports. An AADL event port is the extremity of a channel that exchanges software signals between two entities: the content of the information that goes through this port represents the occurrence of the corresponding signal. Analogically, the information that goes between the electrodes and the pulse generator corresponds to the occurrence of electrical signals either coming from, or heading toward the heart.

Consequently, we have added two event ports at the entrance/exit of each of these two subsystems: electrode and pulse generator. One of these ports will be connected to the electrode implanted in the atrium, while the other will be connected to the electrode implanted in the ventricle. The use of in-ports and out-ports is justified by the fact that each electrode can play the role of a sensor (which then becomes a sender of signals: an out-port at the exit of the "electrode" system) or that of a myocardium stimulator (it receives the command from the pulse generator: an in-port at the entrance of the "electrode" system). The characteristics of the electrical signal that corresponds to the electrodes (impedance, for example) will be described as properties associated with ports of the AADL components that represent these electrodes.

The modeling of the interfaces between the configuration subsystem and the pulse generator needs more work, because the nature of the interactions is more varied. We will limit ourselves to a subset of these interactions, considering that overcoming this restriction does not present any additional complexity: it will suffice to extend the proposed modeling, while using the same concepts as the ones presented here.

We will focus on two types of interactions, namely the ones that enable the physician to:

1) visualize the functioning state of the pulse generator via the supervision interface. Here, we will concentrate on the data collected and measured, in particular on the battery level;

2) modify the configuration of the pulse generator, either by means of a change in the functioning mode (defined in Chapter 2: *O, A, V, D*), or by modifying the characteristics of the pulse generator, such as the minimal pulse rhythm (defined in Chapter 2: *LRL*), or by the minimal delay between a pulse in the atrium and a pulse in the ventricle (defined in Chapter 2: *AV*).

In our AADL modeling, we will group together these interfaces within AADL interfaces groups (*feature group*), which allows us to limit the number of connections between the two subsystems considered. We will then have only one connection between these interface groups (*feature group*) as opposed to a connection per interface without using this concept.

To model the different interfaces of the group, we have to choose from the types of interfaces proposed by the AADL language: *data port, event data port, data access, subprogram access*, etc. We are guided in our choice by the nature of the interaction (data exchange or no data exchange, shared data between different entities, is the interaction crucial or not for the well functioning of the pacemaker and in particular for observing the time characteristics):

– The measurement of the functioning parameters of the pulse generator is executed by the generator but initiated by the supervision entity. We will model this interaction via an AADL subprogram (*subprogram access*). Thus, when the physician needs to know the functioning parameters of the generator, the configuration/supervision station asks for this interface. This triggers a measurement processing from the pulse generator, which sends the result to the entity manipulated by the physician. This access to the subprogram is provided by the generator (which executes the corresponding subprogram) and is required by the supervision entity (which triggers the measurement).

– The changing of AADL modes are triggered by means of event ports. We will therefore add as many event ports as the functioning modes that we wish to model. We decided to identify each port by a name made up of a prefix *to_* plus a functioning mode. Thus, *to_M* will be used to identify an event that triggers a change of mode whose target is the functioning mode "M".

– The interfaces for adjusting the configuration parameters of the pulse generator are represented by data ports, which will allow the transfer of the new configuration values from the configuration station toward the pulse generator. The alteration of these functioning parameters does not cause any particular processing on the pulse generator (besides the modification of a configuration value). Therefore, we do not need to associate an event (that would have triggered the execution of a task) with this particular data port.

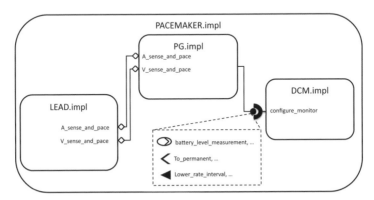

Figure 11.1. *Structure system of a pacemaker using AADL*

Figure 11.1 represents the structure of the pacemaker and its subsystems, as well as the interfaces and connections that allow these subsystems to interact: the main component (*PACEMAKER.impl*) is made up of three subcomponents (*LEAD.impl*, *PG.impl* and *DCM.impl*) that correspond to the electrode subsystem, the pulse generator subsystem and the configuration and supervision station. According to the explanations we have given in this section:

– the pulse generator is connected to the electrodes via two event ports at the entrance and at the exit, corresponding to each of the compartments of the heart: *A_* for *Atrial* and *V_* for *Ventricle*;

– the pulse generating system is connected to the supervision and configuration system via a feature group that has access to several subprograms (*battery_level_measure*, etc.) for measuring the functioning parameters of the generator, the event ports for changing the modes (*to_Permanent*, etc.) and the data ports for modifying the configuration of the generator (*Lower_Rate_Limit*, etc.).

The description of the structure of the pacemaker, which we have presented in section 11.2, also provides the repartition of these subsystems into different, interconnected hardware resources. The following subsection presents the AADL modeling of the corresponding execution platform.

11.3.2. *Execution and communication infrastructure*

The nature of the hardware platform that allows us to model the features of the pacemaker depends on the considered subsystem. The electrode subsystem is made up of different simple hardware peripherals that send an electrical signal to the pulse generator. We will therefore model each electrode by means of an AADL peripheral (the component *device*).

The pacemaker is made up of two electrodes, each of them placed in one of the compartments of the heart. Consequently, we will have two peripherals in the AADL model, one per cardiac compartment. On the other hand, each of these peripherals delivers to the pulse generator the electric signal that corresponds to the heartbeat, whether it is natural or artificial. Therefore, the AADL peripherals will have each event port at the entrance/exit, connected to the port that corresponds to the pulse generator system (*A_sense_and_pace* or *V_sense_and_pace* in Figure 11.1). The peripherals corresponding to the electrodes are then connected to a communication bus that will itself be connected to the execution platform of the pulse generator. This connection represents the hardware support (an electrical wire in this case) that channels the signals between the electrodes and the pulse generator. The latter, as well as the generator configuration and supervision entity are calculators that need a hardware resource capable of executing their features. We therefore model the platform of each of these subsystems using an AADL processor (*processor* component). The two processors thus modeled are connected by a bus that represents the telemeter, which is the communications support between the pulse generator and the configuration and supervision entity.

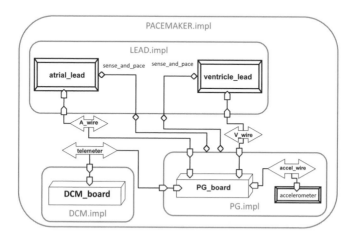

Figure 11.2. *Hardware structure of the pacemaker using AADL*

Figure 11.2 represents the execution platform of the pacemaker. As shown in this figure, the peripherals (*Atrial_Lead* and *Ventricle_Lead*) are defined as subcomponents of the system *LEAD.impl*. Similarly, *PG_board* (pulse generator processor) and *DCM_board* (configuration and supervision entity processor) are, respectively, defined as subcomponents of the *PG.impl* and *DCM.impl* systems.

The event ports of the peripherals are connected to the event ports of the pulse generator. These links allow us to model the interactions between hardware and software components as these two fields are not independent from one another, and the modeling of their dependencies is not limited to the allocation relations between software and hardware components.

Finally, the buses *A_wire*, *V_wire* and *Telemeter* represent the physical communication channels between:

– the electrode placed inside the atrium and the processor of the pulse generator;

– the electrode placed inside the ventricle compartment and the processor of the pulse generator;

– the processor of the pulse generator and the processor of the configuration and supervision entity.

The execution platform thus described is ready to receive the features of the pacemaker, and to help the different software and hardware components interact with each other. Thus, we continue the AADL modeling of the pacemaker by following the second phase of our guidelines (which are presented in section 11.1): the modeling of the software architecture, and the modeling of the behavior of the components of this architecture.

To explain this modeling while still remaining concise, we have chosen to focus on the model of the pulse generator features. This restriction does not mean that we will ignore the interactions between the pulse generator and its environment, but rather that we will not be modeling the software components of the configuration and supervision station. This choice is not without purpose: the pulse generator is obviously the central and crucial element of the pacemaker given that it coordinates the stimulations made by the pacemaker.

We will now enter a new phase of our modeling process: from hereafter, we will focus on designing a software application that responds to the system requirements of the pacemaker. As we have explained in the introduction of this chapter, this requires a good understanding of the links between these requirements and the field of application of the pacemaker, the details of the functioning of the heart as well as knowing precisely what type of cardiac anomaly a given functioning mode responds to. The aim of this learning process is to gain a better understanding of the expected behavior of the pulse generator, and to be able to design the application in a very precise manner. We do not aim for an exhaustive description, but readers can refer to the [BAR 10] handbook for more information. However, before presenting our modeling, we need to explain a certain number of features presented by the pulse generator.

11.4. Overview of the functioning of the pacemaker

In this section, we will do a brief overview of the functionalities of the pacemaker. We will begin by reviewing and specifying the decomposition of the pacemaker into operating modes, then we will describe the functionalities of the pacemaker associated with a subset of these modes. We have chosen to limit ourselves to such a subset because the exhaustive modeling of the pacemaker would be difficult to describe in a single chapter.

11.4.1. *The operational modes of the pacemaker*

The functioning of the pacemaker (or more precisely of the pulse generator) can be decomposed in operational mode. A functioning mode allows us to identify in a synthetic fashion a set of functionalities that a system must provide when used in this functioning mode. The pacemaker presents a significant number of operational modes, organized in a hierarchical fashion: the five modes *permanent*, *temporary*, *pace-now*, *magnet* and *power-on-reset* are the main modes of the pacemaker. In each of these modes, the pacemaker provides a subset of functionalities identified by means of a sub-functioning mode. We will look more particularly into the *permanent* functioning mode because it constitutes the nominal functioning mode of the pulse generator.

11.4.2. *The operational sub-modes of the pacemaker*

In the *permanent* mode, the functioning of the pacemaker is divided into 19 functioning sub-modes in response to the following criteria:

1) What are the simulated cardiac compartments (the atrium, the ventricle, none, both)?

2) What are the investigated cardiac compartments (the atrium, the ventricle, none, both)?

3) What type of response is given to the detection of a heartbeat (triggered, inhibited, tracked, no response)?

4) Is the pulse constant, or is it regulated according to the activities of the patient, detected by an accelerometer?

Sensors and actuators (criteria 1 and 2). The explanation of these two criteria is insignificant, since the issue is mainly to consider which are the sensors and the actuators that the pulse generator has access to in order to regulate cardiac activity.

Type of response (criterion 3). The third criterion corresponds to different types of behavior of the pacemaker:

– When the type of response to the measurements is "none", the stimulations are triggered without any synchronization with the detection of natural heartbeats.

– When the type of response to the measurements is "triggered", the detection of a heartbeat in a compartment immediately triggers a stimulation of the corresponding compartment.

– When the type of response to the measurements is "inhibited", the detection of a heartbeat in a compartment triggers the inhibition of a pending stimulation for that particular compartment.

– When the type of response to the measurements is "tracked", a heartbeat in the atrium must be followed by the stimulation of the ventricle after a fixed delay period called *AV_delay* (delay between a heartbeat in the atrium and a heartbeat in the ventricle), except if, in the meantime, another heartbeat has been detected in the ventricle.

Regulation frequency (criterion 4). The last criterion allows us to modify the manner in which the regulation rhythm is determined by the pacemaker (with or without consideration for the movements of the patient).

The different sub-modes of the pacemaker are identified in the system analysis by means of a suite of four characters that represent, one-by-one, the responses to the first three criteria presented above. For criteria 1 and 2, the possible letters are 0 for none, A for the atrium, V for the ventricle or D for dual. For criterion 3, we have 0 for none, T for triggered, I for inhibited or D for tracked. Finally, the letter R (*Rate*) can be added if the regulating cardiac rhythm depends on the movements of the patient, which are detected by an accelerometer.

For example, VDDR signifies that only the ventricle is stimulated, the two compartments are investigated, the type of response is "tracked", and finally, the movements of the patient are indeed taken into consideration in order to determine the rhythm of the cardiac stimulations.

The types of responses (criterion 3) of the pacemaker mainly depend on the cardiac disease that is being treated. Heart failure ("triggered" response), valvular disease ("inhibited" response or "tracked" response) or arrhythmia ("tracked" response) will guide the configuration of the pacemaker into one or another of these operational mode.

It becomes apparent that the behavior of the pulse generator will be very different in each of these modes. Furthermore, the analysis thus proposed is too imprecise to allow for a detailed design of these functionalities. In order to be able to design these functionalities in more detail, we must specify the behavior of the pulse generator in each of these operational modes. We have chosen to focus on four of these modes (studying all of the 19 modes would be beyond the scope of this chapter).

11.4.3. *Some functionalities of the pacemaker*

The main functionality of the pacemaker is to generate cardiac pulses, which is done via an embedded software. This software is a pulse regulator, that is a control loop that, starting from the measurements of a cardiac pulse, and according to the selected operational mode, emits stimuli toward the muscles of the heart (either toward the ventricular compartment, the atrium or both of them).

As we have already explained, the pacemaker offers a significant number of operational modes. Our aim being to synthesize, we choose to focus on the permanent mode (the nominal mode of the pulse generator), as well as on the following operational modes: DOOR, AAI, VVT and VDD. This allows us to represent all the response types of the pulse generator (0, I, T and D).

In the following, we will describe the functionalities associated with each of these operational modes.

Permanent/DOOR. In this mode, the two chambers (atrium and ventricle) are stimulated, none of them being sensed, and the type of response to the measurements is "none". With this type of response, the pacemaker beats at a rhythm that is independent of the intrinsic pulse of the patient's heart. A pulse in the atrium is separated from a pulse in the ventricle by a delay determined by the atrium–ventricle (AV) period. The rhythm of the heartbeat is determined by the data provided by an accelerometer regarding the physical activity of the patient.

These characteristics are summed up by three timing requirements, which must be observed by the pacemaker:

– the *AV* delays, minimal delay between a heartbeat in the atrium and a heartbeat in the ventricle;

– the *LRL* limit, which imposes a minimal rhythm (or maximal delay) between two heartbeat cycles;

– the *URL* limit, which imposes a maximal rhythm (or minimal delay) between two heartbeat cycles.

This was the main type of response in the first generation of pacemakers. In more modern pacemakers, this type of response is obsolete and only serves to test the pacemaker when a special kind of magnet is placed above it (*magnet* main mode). This mode is also available in the "*permanent* mode" of numerous pacemakers, in order to ensure the retro-compatibility of their functionalities.

Given that the pacemaker sends commands independently of the patient's heart natural heartbeats, there might be a simultaneity between the natural artificial heartbeats induced by the pacemaker. However, the commands do not turn into heartbeats unless they are sent outside of the absolute refractory period of the respective compartment. Furthermore, the risk that an emitted pulse in the vulnerability period of a cardiac compartment should produce fibrillation in the respective compartment is very low; this risk is otherwise related to other risk factors that can be easily identified from a medical point of view.

To further reduce this risk, the types of responses implemented in the latest generations of pacemakers seek to suppress the simultaneity at the level of the heartbeats, which corresponds to the "inhibited" response type.

Permanent/AAI. In this mode, only the atrium is stimulated and sensed. The type of response to the measurements is "inhibited", which means that the pacemaker sends stimuli to the atrium periodically, unless a natural heartbeat is detected before the periodic delay passes. In this case, the internal clock of the pulse generator is reinitialized and the stimuli are only emitted by the pacemaker when the next period of time (fixed by the time limit *LRL*) is reached and on the condition that no natural activity is detected in the meantime.

Here, we may note that the detection of the intrinsic heartbeats of the atrium only starts after a delay that corresponds to the refractory period of the atrium. This is in order to avoid perceiving the electrical residues of a heartbeat as new actual heartbeats.

This operational mode is thus characterized by three time exigences: LRL and URL, which have the same role as before, and ARP, which represents the refractory period of the atrium. Once a heartbeat is detected or triggered in the atrium, new detections are ignored until the delay period is over.

In this mode, there is no longer a simultaneity between the natural artificial heartbeats. On the other hand, it is important to bear in mind that this mode is used on patients whose conductivity between the atrium and the ventricle is normal. Thus, the stimulation of the atrium will trigger a progressive depolarization of the heart and will entail a ventricular heartbeat after the natural AV delay is over. The AV delay is not, therefore, considered in the software design of this operational mode.

Permanent/VVT. In the VVT mode, only the ventricular compartment is sensed and stimulated. More precisely, the measured phenomenon is the depolarization of the interface between the atrium and the ventricle, representative of an atrium pulse followed by the AV delay.

The "tracked" type of response imposes that the ventricular compartment is stimulated as soon as the depolarization of the AV interface is detected. However, the name for this type of responses is misleading because we must also observe the following time constraints:

1) the minimal pulse rhythm (LRL for *Lower_Rate_Limit*), a period at the end of which the ventricular compartment will be stimulated;

2) the refractory period of the ventricle (VRP for *Ventricle_Refactory_Period*), during which the measurements are ignored;

3) the maximal pulse rhythm (URL for *Upper_Rate_Limit*), without considering the measurements unless they are separated by a minimal period.

This operational mode is equally addressed to patients whose AV conductivity is normal. The aim here is to ensure the stimulation rather than the inhibition. This mode is rarely used nowadays, the VVT mode being preferred instead.

Permanent/VDD. In this mode, the two cardiac compartments are sensed, and only the ventricle is stimulated. When the pacemaker detects a heartbeat in the atrium, the pulse generator waits for a delay identified in the system analysis by the AV variable and this corresponds to the AV interval. When this delay starts, there are two possible situations. Either the pacemaker has detected, in the meantime, a natural heartbeat in the ventricular compartment, in which case the pulse generator does nothing, or, in the opposite case, the pulse generator sends a stimulus toward the ventricle.

This operational mode is characterized by four time intervals:

– The AV interval.

– The minimal interval between two pulses (represented by URL).

– The maximum interval between two pulses (represented by LRL).

– The post-ventricular refractory period (PVRP) of the atrium, which allows us to avoid mistaking an electrical residue coming from a ventricular heartbeat for an actual new heartbeat.

Now that we have detailed these features, we will detail their modeling using AADL language.

11.5. AADL modeling of the software architecture of the pulse generator

The modeling of the features of the pulse generator uses the following AADL components: a *process*, which represents the program deployed on the execution platform of the generator, and several tasks (*thread*), which represent the main features of the program, that is the control loop, the configuration and the supervision of the system.

11.5.1. *AADL modeling of the operational modes of the pulse generator*

In the AADL model, we shall enumerate the set of operational modes of the pacemaker. We will thus have five main modes, with which we associate a certain number of sub-modes: 19 in the *permanent* mode, four in the *temporary* mode and one on the *power-on-reset* mode. Each of these functioning modes corresponds to a different subset of functionalities, which we can represent in AADL due to the different variations of the pulse generator's software architecture. For this purpose, the set of the operational mode in which this entity remains accessible can be associated with a model entity.

To represent the set of modes in an AADL model, we have several possibilities. Let us start with a poor, if classic, solution in computer science: flattening the hierarchical structure. We could enumerate the set of modes (29 in total), identifying them by the concatenation between the name of the main mode and the series of letters identifying the sub-mode. We will thus have the following modes: *permanent_Off, permanent_DDDR, permanent_VDDR, ..., temporary_OVO, temporary_OAO, ..., pace-now_VVI, magnet_AOO, magnet_VOO, ..., power-on-reset_VVI*. Because we aim to provide a synthesis, we will not describe the exhaustive list of the operational mode. This can be deduced from the system analysis of the pacemaker [BOS 07]. The pulse generator's features vary depending on the functioning mode in use, and in order to represent these variations, it suffices to associate the set of the operational mode with the main task of the generator (*thread* component), and then to associate a different behavior with this task in each of the modes.

We have adopted a different solution that seeks to preserve the hierarchical structure of the modes. We will thus obtain a more legible model, and hence a model that is more easy to maintain. To do this, we will associate the five main modes of the pacemaker with the control process (*process* component). We will then add an implementation of the task (*thread implementation* component) per functioning mode, and then we will associate with each task the functioning sub-modes of the main mode that this task represents. Finally, we will associate a different behavior with this task in each of these sub-modes.

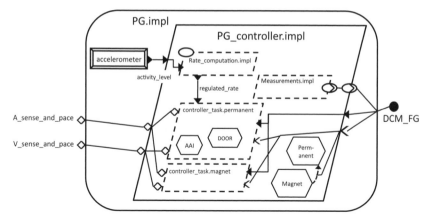

Figure 11.3. *Software structure of the pacemaker using AADL*

Figure 11.3 represents the AADL architecture of the pulse generator. This figure shows four important parts of the functioning of the pulse generator:

– The management of the operational mode: as we have explained before, we have chosen to represent the main modes as modes in the *PG_controller.impl* process component. We have represented two of these modes (*Permanent* and *Magnet*) in Figure 11.3. The event ports of the feature group (*DCM_FG*) connected to the configuration station manages the changing of modes.

– The management of functioning sub-modes: the functioning sub-modes are associated with each of the tasks that correspond to a main mode. For the regulation task of the heartbeats in the *Permanent* mode (*controller_task.permanent*), we have represented two of these modes (*AAI* and *DOOR*) in Figure 11.3. The management of sub-mode changing is done equally by means of the interface group (*DCM_FG*) connected to the configuration station.

– The regulation of the heartbeats according to the physical activity of the patient: an accelerometer, connected to a periodic task (*Rate_computation.impl*), provides the data that corresponds to the physical activity of the patient. The task *Rate_computation.impl* converts these data in the heartbeat period that the regulation task must observe (*controller_task.permanent*).

– The measurement of the pacemaker parameters: a task *background* (*Measurement.impl*) is added to the process *PG_controller.impl*. This task, activated upon receipt of a measurement request by an access interface to a subprogram, allows us to measure the functioning parameters of the pulse generator, and to send them to the configuration and supervision station.

Let us emphasize here the fact that a mode changing can alter the structure of the software architecture. For instance, the logical connection between the out-port of the electrode placed inside the atrium and the input port that corresponds to the pulse generator will not be accessible except in sub-modes type XAXX and XDXX. We are dealing here with a software connection/disconnection; the non-accessibility will be implemented using a software ignoring the information that comes from this port. The same goes for the connection between the accelerometer and the pulse generator for the modes XXXR or XXX.

The rest of this section details the behavior of the control task associated with the permanent mode, in the sub-modes we have chosen to study. The variation in the type of responses associated with each sub-mode needs a finer modeling of the behavior of the software application.

11.5.2. AADL modeling of the features of the pulse generator in the permanent mode

We have modeled the functioning of the control loop of the pulse generator with the help of a single task associated with the permanent mode. This task (*controller_task.permanent*) is represented, although partially, in Figure 11.3.

In this section, we will focus on the functioning of this task and on its complete modeling in AADL. Let us note that the behavior associated with this task will be different in each of the sub-modes of the permanent mode. To cover a wide modeling specter while remaining concise, we have chosen to model the following operational sub-modes: DOOR, AAI, VDD and VVT.

Before presenting the behavior of the control task in each of these sub-modes, let us add to its description the elements we will be needing for the rest of this section.

We can first and foremost specify the type of dispatch protocol (*Dispatch_Protocol property* in AADL) associated with *thread controller_task.permanent*. At first sight, the AADL modeling of the control task of the pacemaker is insignificant: it has a periodic functioning and suffices to use a periodic task (*Dispatch_Protocol => periodic*). On a closer inspection, we realize that we must take into account a certain number of additional time properties, such as: a heartbeat in the atrium must be followed by a delay corresponding to the AV interval. In order to avoid wasting our resources, a passive wait is mandatory: the task releases the computing power (processor) and will be reactivated when the AV delay has passed. This means that the behavior of the task considered is not purely periodic since the task is deactivated and reactivated in the middle of the staging. Furthermore, the reactivation interval is programmable and can vary (since it depends on the measurements of the accelerometer).

What types of AADL tasks can we use? The task we wish to model is reactivated upon events that correspond to an internal clock, and the global reactivation period is not fixed. Consequently, the reactivation rhythm is not constant and the *Periodic*, *Hybrid* and *Timed* AADL tasks must be excluded. A sporadic task could fit our purposes, but we do not have the constant minimal delay between two reactivations of the task. We will therefore opt for an aperiodic task and will model its internal behavior with the behavioral annex of the AADL.

Then, we will also add data ports to the *controller_task thread*. These ports will allow us to configure the time parameters of the control loop; the name of each port corresponds to the acronym of the parameter:

– The low stimulation limit (LRL data port).

– The high stimulation limit (URL data port).

– The atrio-ventricular delay (AV data port).

– The refractory period of the atrium (ARP data port), or the post ventricular refectory period (PVRP data port).

Let us equally note that the task *controller_task* has a *rate_value* port (see Figure 11.3), whose values come from the processing of the information provided by the accelerometer.

Finally, we model the operational mode of the control task, as well as the mode changes. To do this, we add four modes to the *thread* AADL *controller_task.permanent*: DOOR, AAI, VDD and VVT. We then define four event ports (*to_DOOR*, *to_AAI*, *to_VDD* and *to_VVT*) and 12 transition modes triggered by the reception of an event on one of these ports: for instance, *DOOR -[to_AAI] –> AAI* defines the mode change from *DOOR* to *AAI* upon reception of an event on the *to_AAI* port of the task. It is thus easy to deduce the content of the other 11 mode changes.

To model the behavior of the task in each mode, we add a behavioral automaton per task mode. We will specify this in the AADL model by adding the «in modes» clause at the end of the definition of the automaton. For instance, the following AADL description corresponds to the behavior of the control task in the AAI mode (see listing below).

```
thread implementation controller_task.permanent
    ...
    annex behavior_specification
4   {**
    ...
    **} in modes AAI;
    ...
end controller_task.permanent;
```

Listing 11.1. *Behavioral appendix associated with a mode*

In the rest of this section, we present the content of the behavioral annexes that correspond to the behavior of the control task in each of these sub-modes that we have chosen to model. The behavioral annex allows us to represent the behavior of the AADL components in the shape of automatons with state transitions. We will represent these automatons with the help of figures where the states will be represented by circles and the transitions by arrows. The conditions and the actions associated with each of these transitions will respect the syntax of the behavioral annex. Furthermore, although the behavioral annex allows us to represent different types of states (*initial*, *final*, *complete* and *execution*), we will only use two of them:

1) An *initial* state, represented by a full circle. To improve the readability of the models provided, we will ignore here the initialization phases of the pulse generator, these stages being implemented when the transition from the initial state of the behavioral automaton takes place.

2) The *complete* states represent the different stages of the behavior of the task. Let us remember here that a *complete* state corresponds to a state were the execution of the corresponding task is put on hold. As we will see in the rest of this section, the functioning of the control task mainly consists of synchronizing the arrival and the emission of different types of events with the time exigences that impose a certain number of delays (the task being put on hold).

Permanent/DOOR. The behavioral automaton represented in Figure 11.4 showcases the behavior of the *controller_task.permanent* task in the DOOR sub-mode. In this sub-mode, the behavior of the control task is relatively simple: after the task is initialized, represented in the figure by the transition starting from the full circle, the task reaches the *Wait_rate_timeout* state where its activity is suspended. The task remains in this state until it is reactivated by the end of the delay dictated by the value at the *rate_value* data ports. The *thread* is thus reactivated, it sends an event to the *asp* port (short for *A_sense_and_pace*) and reaches the *Wait_AV_Delay* state. The activity of the task is again put on hold when it reaches this state, and it will be reactivated when the AV delay comes to an end. Then the task will send an event to the *vsp* port (short for *V_sense_and_pace*) and reach the *Wait_URL* state, where it waits for the necessary time to pass, thus observing the URL time constraint. The task then returns to the *Wait_rate_timeout* state. Let us note that the LRL characteristic is guaranteed as long as the *rate_value* represents the reactivation intervals that are smaller than those dictated by LRL.

Permanent/AAI. The main difference between the *Permanent/AAI* mode and the *Permanent/DOOR* mode is the inhibition and the respect for the refractory period of the atrium. In the AAI mode, the inhibition is modeled as follows: the internal clock of the pulse generating task is reinitialized when a natural heartbeat is detected in the atrium. Respecting the refractory period refers to allowing the detection of a new heartbeat in the atrium only after a fixed delay interval has passed since the previous heartbeat (natural or artificial), an interval that corresponds to the refractory period of the atrium (ARP).

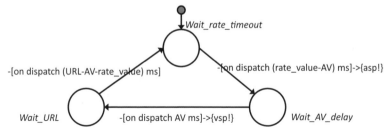

Figure 11.4. *Behavioral automaton of the pulse generator in the DOOR sub-mode*

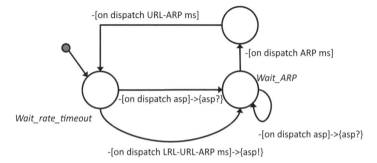

Figure 11.5. *Behavioral automaton of the pulse generator in the AAI sub-mode*

Figure 11.5 represents the behavior described above by using the constructions provided by the AADL behavior annex. When the pulse generator is activated in the *AAI* mode, it reaches the *Wait_rate_timeout* state, where the generation task is inactive. Two transitions allow the automaton to come out of this state, to activate the task and to reach a new synch state called *Wait_ARP*:

– Upon reception of an event at the *asp* port (for *A_sense_and_pace*), the received event is then consumed (see *asp?* in Figure 11.5).

– Upon the end of the delay period LRL–ARP, a pulse is sent on the *asp* port (*see asp!* in Figure 11.5) that corresponds to the port *A_sense_and_pace*.

In this new state of inactivity, the task waits for the ARP delay period to pass: transition toward the *Wait_rate_timeout* state with our condition *[on dispatch ARP ms]*. If the pulse generator, in the *Wait_ARP* state, receives an event of pulse detection in the atrium, it is consumed and the generation task remains in the *Wait_ARP* state: see, in Figure 11.5, the transition reentering the delay state ARP.

This behavior fixes very well the functioning of the pacemaker in the wider AAI mode: the pulse generator only stimulates the heart if no other natural pulse has been detected before the wait time has passed; furthermore, the pulse detected during the refractory period of the atrium is being ignored.

The constraint regarding the minimal pulse rhythm is respected by using the limit LRL–ARP: when the automaton enters in the *Wait_rate_timeout* state, the rhythm flows at more ARP ms than the last heartbeat in the atrium. By consequently stimulating the heart at more LRL–ARP ms, we ensure that the heartbeats take place at least every LRL ms.

Permanent/VVT. In the *Permanent/VVT* mode, as soon as a pulse is detected in the ventricle, the pulse generator stimulates this same compartment. However, in this mode, the pulse generator must observe three time characteristics:

– the minimal pulse rhythm (LRL), a period at the end of which the ventricular compartment will be stimulated;

– the refractory period of the ventricle (VRP), during which the detected pulse is ignored;

– the maximal pulse rhythm (URL), where we do not consider the measurements unless they are separated by a minimal period.

In this mode, we could consider the opportunity for modeling the control task by a sporadic task, since the maximal rhythm of a pulse puts an upper limit on the time interval that separates two reactivations of the said task. Again, this characteristic is not constant. Consequently, we will continue to model the control task via an aperiodic task.

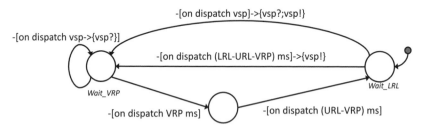

Figure 11.6. *Behavioral automaton of the pulse generator in the VVT sub-mode*

Figure 11.6 represents the behavior of this pulse generation task in the VVT sub-mode. In the same way as the previous automata, the states represented in this automaton are the synchronization states: the task is inactive and waits to be reactivated either by detecting a pulse in the ventricle (transitions whose condition is

-*[on dispatch vsp]*), or by coming to the end of one of the waiting times characteristic of the pacemaker (transitions -*[on dispatch X ms]* where *X* is equal to a mathematical operation performed on the values LRL, VRP and/or URL).

The behavioral automaton of the pacemaker in the VVT sub-mode is much more complex than the automaton presented in Figure 11.6, because the number of synchronizations that must be taken into account has increased. We must first ensure that the minimal pulse rhythm specified by the value LRL is respected. Given the timing characteristics of the pulse generator in mode VVT (LRL, VRP, and URL contraints), a pulse cycle takes at most the URL+VRP ms rate throughout a pulse cycle. The condition for leaving the maximal delay between two heartbeats (*Wait_LRL*) is therefore -*[on dispatch (LRL-URL-VRP) ms]*. This transition is inhibited if, in the meantime, the pacemaker has detected a natural heartbeat. In this latter case, the ventricle is stimulated (which corresponds to the expected behavior: the detection of a heartbeat in the ventricle triggers the stimulation of the latter).

Once one of these two transitions is completed, the automaton waits for the refractory period of the ventricle to end, during which the heartbeats are ignored. The task then waits for the delay that corresponds to the maximal pulse rhythm (URL) to be reached. When this delay begins, we return to the *Wait_LRL* state.

Permanent/VDD. In the *Permanent/VDD* mode, only the ventricle can be stimulated by the pulse generator, when the two chambers are probed. In this mode, the functioning of the pacemaker combines the behavior associated with the VVI mode with the time constraint AV. To implement this mode, we need to consider four time characteristics:

– the minimal pulse rhythm (LRL), a period of time at the end of which the ventricular compartment will be stimulated;

– the PVRP, during which the heartbeats in the ventricle are ignored;

– the maximal pulse rhythm (URL), without considering the measurements unless they are separated by a minimal period of time;

– the AV delay, a minimal period of time that separates a pulse in the atrium from a pulse in the ventricle.

Figure 11.7 represents the behavior of this pulse generation task in the VDD sub-mode.

11.6. Modeling of the deployment of the pacemaker

To finalize the AADL modeling of the pacemaker, it suffices to model the allocation of the software components on the execution platform: the third and final phase of our modeling endeavor. More specifically, we need to associate the logical connections (between the tasks, between the processes, between subprograms, etc.) with the physical connections (communication bus), the software components (processes, tasks, etc.) with the calculation resources (associated memory processes) and the passive components (data) with the storage resources (memory).

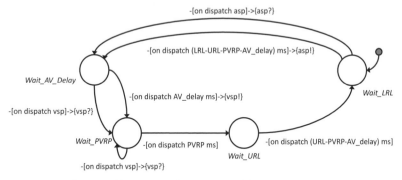

Figure 11.7. *Behavioral automaton of the pulse generator in the VDD sub-mode*

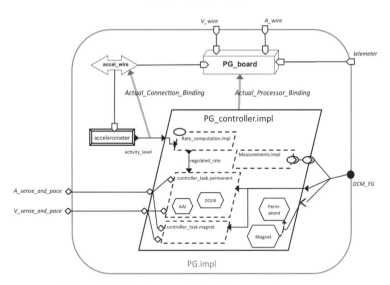

Figure 11.8. *Deployment of the pulse generation process*

Figure 11.8 illustrates the deployment of the regulation process of the heartbeats (*PG_Controller.impl*) on the electronic board *PG_board.impl* associated with the subsystem *PG.impl*. This association, represented by a double arrow, is expressed in textual AADL with the property *Actual_Processor_Binding*. By means of this association, we specify that the calculations related to the regulation of the pulsations will be done on the board *PG_board.impl*.

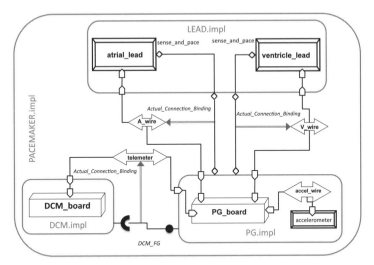

Figure 11.9. *Deployment of the connections between the subsystems of a pacemaker*

In the same way as before, Figure 11.9 illustrates the association of the connections between the different subsystems (pulse generator, configuration and supervision station as well as electrodes) with the communication media that connect these different subsystems (telemeters and electrical wires).

We have thus obtained a complete and relatively precise modeling of the architecture of the pacemaker. Depending on the subparts of the architecture, this work required between three and five iterations. Several iterations would still be necessary for obtaining an exhaustive modeling of the system. It is important to note here that the analysis of this modeling helps us put into practice this process because it allows us to detect the design faults even before the process of producing the system has started. The exploitation of the models described in this chapter is presented in Chapters 12 and 13.

11.7. Conclusion

In this chapter, we have proposed an AADL model of the pacemaker, resulting from various modeling iterations and from the model analysis. This chapter showcases the difficulties inherent to the modeling of a system, and proposes a few suggestions for solving these difficulties. The benefits of this modeling work will be presented in the following chapters, which deal with analysis and code generation based on AADL models.

The model we have provided here is an architectural analysis of the pacemaker, and it includes:

– a description of the expected functioning mode;

– a description of the behavior of the pacemaker in some of its operational modes.

The following chapters will show how to use this model in order to:

– verify that the architecture thus represented observes the manufacturing requirements of a pacemaker (see Chapter 12);

– carry out an implementation that respects this architectural analysis (see Chapter 13).

It is worth noting that the models proposed here are a specification of the pacemaker and do not have any bearing on its implementation, other than having to observe the constraints imposed by the model (e.g. in terms of behavior and allocation). For example, the presence of tasks imposes that certain processes be carried out while delays are respected, but does not impose that these tasks be implemented by the *system threads* of an operating system.

11.8. Bibliography

[BAR 10] BAROLD S., STROOBANDT R., SINNAEVE A., *Cardiac Pacemakers and Resynchronization Step by Step: An Illustrated Guide*, John Wiley & Sons, Hoboken, NJ, 2010.

[BOS 07] BOSTON SCIENTIFIC, PACEMAKER System Specification, January 2007.

Chapter 12

Model-Based Analysis

12.1. Introduction

In this chapter, we will analyze the pacemaker model we have built in Architecture Analysis and Design Language (AADL) in Chapter 11.

This analysis consists of verifying the coherence of the implementation model, as well as part of the system's requirements defined at the beginning of the design. There are several types of analyses that can be carried out, but we will only focus on the requirements regarding the time constraints 22/P and 23/P.

The AADL language belongs to a family of architectural languages having a rich semantic that corresponds to a specific field (i.e. embedded software engineering). We may thus speak of a *Domain-Specific Modeling Language* (DSML).The implementation of the model analysis varies according to the degree of dependence between the language and the implementation of the method used for analysis. Following are three possible situations:

– The analysis tool was developed specifically for analyzing the models of a dedicated language. Thus, the REAL constraint checker [RUB 11] was built with the only goal of analyzing the static properties expressed in the architecture of the system.

– The semantic of the model is extracted and fed into the domain-specific tools (e.g. the scheduling analysis tools [SIN 05]). These tools allow us to efficiently validate a clearly identified subset of system requirements, namely, in this case they allow us to check the response time of the pacemaker.

Chapter written by Thomas Robert and Jérôme Hugues.

– The semantic of the model is extracted and fed into generic verification tools (logic tester, *Computation Tree Logic* (CTL) model checker, reachability analysis tools for automata, or Petri nets).

In the last two cases, the DSML semantic is extracted and passed over to the monitoring tools. Every verification task confronts a model to a requirement. Therefore, we must translate the system requirements into the language of the constraints of the verification tool. Second, we must extract the fraction to be analyzed from the AADL model, formed of a subset of components, connections and properties of the model that is relevant for our analysis:

The rest of the chapter illustrates the following points:

– The validation of requirements 22/P and 23/P for each of the modes.

– The validation of requirement 23/P via a mode change.

For our purposes, we have selected the UPPAAL tool, which has already been introduced in this book. Rather than presenting an existing transformation tool, we have chosen to present the process of model construction, so as to introduce the reader to the different stages enabling them to link the analysis of an AADL model with the verification of an UPPAAL model.

12.2. Behavioral validation, per mode and global

This section details the phase where the semantic of the behavioral automata is extracted and translated into UPPAAL. The validation of the requirements (22)/P and (23)/P serves to showcase the overall aim of the procedure, its advantages and its disadvantages.

This discussion will help us illustrate the problems regarding the extraction phase, a phase that must be undergone as soon as there is a discrepancy between the modeling formalism and the verification formalism. There are other gateways between AADL and the behavioral verification formalisms [ABD 08, CHK 08, BER 09, BOZ 10][1]. All of these gateways follow the same procedure, with different degrees of coverage of the input model. Depending on the tools, the complexity of the model can be limited: a distributed monoprocessor, distributed carefully considering the hardware platform. It is thus useful to know the modeling hypotheses of the tool well, that is of the models it knows how to handle.

The AADL conversion toward UPPAAL will cover the elements that UPPAAL can process validly, and the properties that are of interest to us. It will exclude the

---

1 Check the website of the AADL committee for a more exhaustive list of tools: www.aadl.info

properties and the elements that UPPAAL cannot express. Here, we will be more particularly interested in the behavior of the system tasks, their interaction and their temporal behavior. We will therefore restrict our analysis to only the components of the model that have an impact on these behavioral properties.

12.2.1. *Validation context and fine tuning of the requirements*

The modes, the behavioral automata and tasks have a semantic that is similar to that of communicating automata networks. This claim is backed by the semantic of port connections between tasks. However, such a formalism does not allow us to benefit from information such as the latencies of message propagation, or to have access to a common time base. Furthermore, the analysis of such models introduces several decidability problems [BRA 83] – that is, whether we can build an algorithm that is able to provide an answer to the question posed by a chain of elementary actions.

It is therefore important to use concurrency models for which the accessibility analysis of a state is decidable. This property is necessary as soon as we seek to verify the "safety" properties, by analyzing models[ALP 87, KUP 99].

The translation in UPPAAL was chosen for the simplicity of its model of concurrency and its capacity to model the conditional transitions that bind the time flow (transitions that are only active under certain conditions). Finally, the model is decidable and a complete tool support is offered. A complete tutorial for carrying out the formal verification under UPPAAL is available on the tools' website [BEH 04].

12.2.2. *Translation of the behavioral automata into UPPAAL*

UPPAAL enables the verification of the properties expressed in temporal logic (CTL) on a network of timed automata [ALU 99] extended with bounded integer variables. The state space of the formulas (CTL) is made up of Boolean variables and bound integers.

12.2.2.1. *Timed automata*

A timed automaton is made out of discrete states called locations and conditional transitions, which connect these states:

– the state of an automaton is characterized by the location it occupies, and by a set of "real" variables that correspond to clocks. These variable values increase homogenously, to model time elapsing, and can be reinitialized in order to measure how much time has passed since a given event. These clocks are used for defining the necessary conditions to fire the transitions. UPPAAL extends this model using bounded integers. The latter vary when a transition is undergone. The modification can be expressed using an assignment operator and a simple language made of arithmetic expressions;

– the transitions can indeed be implemented provided that the state of the automaton (the location occupied, the clocks and the integer variable) validates a simple logical condition. This condition is called a guard. The guard of a transition can be trivial (always true) or a conjunction of elementary constraints (e.g. $variable < Constant$);

– UPPAAL allows us to constrain the manner in which time passes in a "location". This may range from: "the time cannot progress", representing the instantaneous situations of concatenation (locations called $Urgent$), to "clock x must stay inferior to $Constant$" for expressing the fact that an action must necessarily be executed before reaching this value.

Figure 12.1 depicts the automaton representing the behavior of the *controller_task* in DOOR mode.

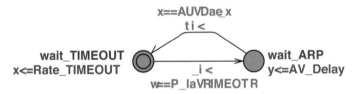

Figure 12.1. *Timed automaton equivalent to the AADL automaton for DOOR 11.4*

This model defines two locations: `wait_TIMEOUT` and `wait_ARP`. Besides the location, two clock variables are defined: `x` and `y`. The automaton stays in location $Wait\_TIMEOUT$ as long as the clock satisfies $x < Rate\_TIMEOUT$, and changes to location `wait_ARP` when the clock reaches the instant $Rate\_TIMEOUT$. The signal ap is then emitted.

12.2.2.2. *Definition of the translation problem*

In what follows, we wish to provide a solution for the problem of translating AADL models into UPPAAL automata. In other words, we seek to implement the scheme described in Figure 12.2.

The figure illustrates the process of translating AADL modes and the system requirements into UPPAAL in order to carry out – using UPPAAL – a verification "equivalent" to the one that could have been carried out directly on the behavioral automata of AADL language. It only makes sense to use this approach if a false property in AADL is necessarily false in UPPAAL as well. We can then talk of process correction. Conversely, if a property is valid in AADL, it should not give a contradictory verdict in UPPAAL. This second aspect allows us to define the precision of the approach.

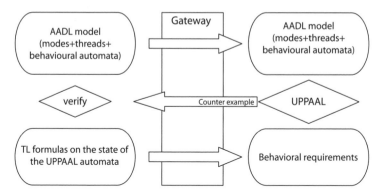

Figure 12.2. *Translation of the models from AADL into UPPAAL and vice versa*

Thus, if we wish to be able to define the conversion into UPPAAL, we need to solve two problems:

– defining the semantic of the AADL behavioral semantic automata as state/transition automata;

– "translating" the scenarios we wish to assess into CTL predicates, used by UPPAAL in order to express the requirements we must verify.

One important point is to guarantee the traceability between AADL models and UPPAAL models in order to ensure the equivalence between models. This is most often done via conventions regarding the denomination of the generated entities.

12.2.2.3. *Semantic of the behavioral automata*

The validation of the requirements 22/P and 23/P first goes through the analysis of the model's elements that define the behavior of the *controller_task* in the long term. The behavior of a task depends on the following elements:

– Concurrency/scheduling model of the tasks and of the state of the scheduler.

– Intertask communication model and the state of the communication buffers.

– Synchronization between the communication ports and the tasks.

– Synchronization between the clocks and the behavior of the tasks.

– Impact of the modes and the modes' transitions.

The question becomes: "How can we integrate all of this information into an UPPAAL system? Must we create a description while considering all of these elements?" In what follows, we will provide an overview of all of these different problems and indicate appropriate response elements.

The scheduling: in a first stage, we will eliminate the question of the modeling of task scheduling. These tasks are characterized by attributes that define their periods, their relations of priority and causality. We will focus on validating the coherence of these parameters in relation to high-level constraints. The feasibility of the scheduling and deployment is outside the scope of the translation into UPPAAL.

If we wish to consider the scheduling, all we need to do is model the intervals where the task can emit stimulation events. This allows us to prove that the behavior of the task verifies the high-level requirements.

The ports and the synchronization models: the AADL communications have an impact on the behavior of the set of active components (tasks). There are different synchronization levels between the "read/write" operations on a port and the actual events of sending and receiving, as well as the behaviors associated with the reception of these data. These two events can be largely out of synch. It is therefore useful to model this potentially complex synchronization model.

There are three elements that need to be modeled in order to represent a communication port: the waiting queue (which is size one, by default), the management protocol of the overruns in the queue and the synchronization with the tasks. We will make the following choices:

– we will consider the *dropNewest* type of overflow management protocol. Where there is queue overflow, the new messages will be considered to be lost and an error event will be generated;

– the synchronization with the tasks corresponds to the triggering of an activity (which introduces a causality relation between the two events);

– the modeling of the ports will be made via the definition of an automaton serving to represent the behavior of the communication port. This automaton will be synchronized with the automaton of the task to which the port belongs. An additional synchronization will be necessary in order to represent the causality relation that corresponds to the sending and receiving of a message.

Out of these choices, we will deduce the UPPAAL modeling patterns:

– a waiting queue is modeled by representing each padding state of the waiting queue. In the standard case, there are two states corresponding to the following conditions: "empty queue" and a "current event";

– the management protocol of the buffer overflow is taken into account by adding transitions to previously defined padding states. They model the case where an extra message is recieved and cannot be stored, entailing the production of "error" event;

– the content of a port is accessed by adding a specific event that allows us to synchronize the port and the task automata. We will note this event using the name of the port followed by "_in" for the in-ports.

In the case of the *controller_task*, we thus obtain four automata. The automaton in Figure 12.3 represents one of the automata of the event port.

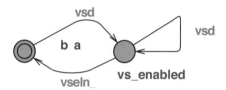

Figure 12.3. *Model of a port and its synchronization events*

Currently, we must handle the situation where data are transmitted through ports. We hereby encounter the first obstacle in the use of UPPAAL for realizing the verification of the system. These data ports are used either for representing configuration constants, or values that vary throughout the time. We will focus on the case where only the configuration constants are used, for instance in the VVT and VOO modes. The table on page 28 of the analysis of the pacemaker allows us to understand which parameters actually play a role in each mode. In this case, we will represent these values as constants transformed into parameters at the time of the definition of the UPPAAL timed automata. This will allow us to model, simply enough, a dispatch upon "timeout" in the case of a hybrid task (which is a case not covered by the existing gateways). We make this modeling choice because these parameters are used in guards that involve clocks and cannot change throughout the time due to timed automata formalism limitations.

The table that specifies the authorized variation intervals for each parameter allows us to carry out scripts that define the different values for TIMEOUT and SAFETY_DELAY, as well as ventricular refractory period (VRP). These scripts enable us to create an automaton for each configuration of the parameters.

The activation conditions this stage of transformation into UPPAAL is based on the interpretation of the states of the AADL behavioral automata called "complete state" and of the transitions with "dispatch condition".

We may distinguish between following two activation (or *dispatch*) conditions:

– A condition upon the expiry of a deadline that is relative to the moment when the automaton goes into the final state.

– A condition upon the presence of an element in the queue of an event port.

In the first case, we will apply the transformation described in Figure 12.4. The projection proposed is based on a translation of different unitary elements:

– the emission of an `action` event at the AADL level will be translated into an UPPAAL event $action\_out$;

– the dispatch condition will be translated into UPPAAL in: (1) an invariant forcing a transition at the moment when the delay expires $X = Y$, (2) a guard disabling the dispatch transition, to avoid firing the transition before the expected time $X <= Y$ and (3) the reset of a clock used to measure the time passed in complete state. This clock measures the delay between entering the state and the moment at which the activity restarts, that is $X = 0$.

Let us note that a single clock is necessary for processing the set of this type of transitions for each automaton. This is an important point: a huge number of clocks makes the UPPAAL accessibility analysis even more complex.

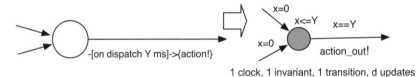

1 clock, 1 invariant, 1 transition, d updates

Figure 12.4. *Transformation of a dispatch transition on "timeout"*

In the second case, which corresponds to dispatch sources connected to event ports, we will use transitions synchronized with the automata representing the in-ports or the out-ports. Figure 12.5 represents the automaton of the control task and its port automata. We should note the combination between the use of the expansion of a dispatch triggered by the time (zone 1), and the dispatches triggered on an event (zone 2).

12.2.3. *Refining requirements 22-23/P*

The first phase in a validation task is to avoid as much as possible the ambiguity found in the requirements that define the expected behavior of the system. Requirements 22 and 23 use different concepts without necessarily defining them and introduce a certain freedom of interpretation. First and foremost, let us review the definitions:

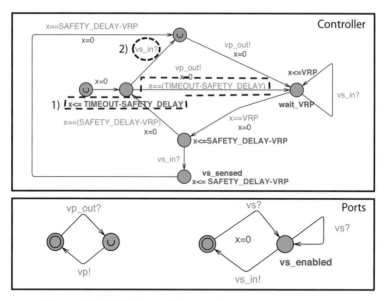

Figure 12.5. *UPPAAL model of a task and its ports (VVT mode)*

– LRL constraint (review): in modes DXX and VXX, the lower stimulation limit (or *lower rate limit* (LRL) starts when a pulse is sensed in the ventricle, or when a stimulation is detected and still corresponds to a minimal rate of pulses per minute.

– URL constraint (review): the upper pulse rate *upper rate limit* (URL) is the maximum number of events measured at the level of the atrium. Somewhat dualistically, the URL is the minimum interval between an event in the ventricle and the next stimulation of the ventricle.

The constraints on the pulse rates can be interpreted in terms of minimal/maximal delays between two events. This approximation is a correct interpretation from the safety point of view. We are dealing with a reinterpretation of a mathematical property.

In what follows, we will use TIMEOUT to designate the maximum authorized delay, that is a deadline. This quantity allows us to express the sufficient condition for the 22/P requirement, and SAFETY_DELAY – the minimum duration between two stimulation events on the leads of the pacemaker, allowing us to express the sufficient condition to satisfy 23/P.

This allows us to refine these two requirements (we add the "R" suffix in order to identify the refining and to trace the heritage that exists between 22/P and 22/P/R):

– 22/P/R: the interval between two stimulation events (natural and detected, or artificial and created) must be inferior to TIMEOUT;

– 23/P/R: the minimum duration between an (artificial) stimulation event and the previous artificial or natural stimulation event must be greater than SAFETY_DELAY for the same element of the heart (atrium or ventricle).

12.2.4. *Study of the permanent/VVT mode*

By applying principles we have described above, we have obtained the automaton of the permanent/VVT mode as presented in Figure 12.5.

We must still express the constraints that serve to verify the requirements 22/P/R and 23/P/R in CTL. We have chosen the same method as the one proposed for the analysis using MARTE: the formal observers.

The observers resulting from these requirements are quite simple (Figure 12.6). These requirements monitor the reception and the emission of events on the system electrodes. For 22/P, we loop back on the same state, and we reset the clock to zero upon reception of the event vs or vp. If no event is detected at the end of TIMEOUT, then we enter a state of error. The automaton for 23/P follows the same logic, only the guards change: an error is detected if vp is received before the minimal SAFETY_DELAY. The clock y that measures the arrival of vp events is reset to zero as soon as the events are sufficiently far apart in time.

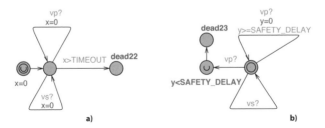

Figure 12.6. *Observers: (a) requirement 22/P for VVT, (b) requirement 23/P*

12.2.4.1. *Complexity and combinatorial explosion*

To verify the set of acceptable configurations of the VVT mode, a script allocates the values of the TIMEOUT, VRP and SAFETY_DELAY constraints, and then starts the verification. We then proceed to a decomposition of time into discrete units, in order to test different configurations. The verification of each configuration takes between 5 and 10 sec. We stumble upon the problem of combinatorial explosion, however, because the number of possible configurations is in the order of 10,000. This number is very high, compared to the relative simplicity of the models considered.

12.2.5. *Study of the changing of the permanent/VVT→Magnet/VOO mode*

The AADL model in section 11.4.1 defines the mode switches that are allowed for the impulse generator. Among other things, the modes allow us to define which of the active structures can be effectively executed on the hardware at a given time. This is the abstraction that has been used to separate the different behavioral models deduced from the operating modes of the pacemaker (DXX, VXX, etc.).

12.2.5.1. *Mode switch and requirements*

The model considered so far represented the behavior of the impulse generation task in the VVT functioning mode. We will focus on the mode changing problem associated with the control task of the impulse generation. Each mode switch has a destination mode that has a different task model associated with it (at least as far as the behavior description clause is concerned).

By default, a task can only switch modes when already in "complete state". Each mode switch is associated with an event port that allows for the transition to be triggered. A task activated due to a mode switch always has the same activation procedure. If a state must be transferred during the mode switch, we must manage it before sending out the event toward the mode changing port.

When the transition is done from the Permanent mode to the magnet mode (at the level of the process that represents the pacemaker), the control task must pass from the VVT mode to the VOO mode. We will verify if the LRL constraint remains valid despite the mode switch.

The behavioral automata proposed in the model only contain the "complete state". By default, the mode switch is thus allowed in each of these states. A set of transitions is added to the existing model in order to represent the mode switch: one transition per state that is equivalent to a "complete state". These transitions allow us to suspend the activity of the model associated with the VVT mode, as a consequence of the mode switch modeled through the reception of the *to_VOO* event. By default, the mode switch can start from a complete state, only when the associated event port contains an element. Upon the activation of the destination mode, the associated task model resumes its activity from the initial state.

12.2.5.2. *Verification and assessment*

We have built the two corresponding models and demonstrated the possible violation of requirement (22)/P and (23)/P throughout the mode changing process. This fault comes from the absence, in the current version of the model, of a mechanism that allows us to preserve, throughout the mode switch, the memory of the last dispatch time. However, this information is necessary for controlling the transitions in the VOO mode (in order to be able to respect the 22/P requirement as well as the 23/P one).

The execution counter examples place us in situations where the mode switch entails incorrect *dispatch* times. The following two scenarios may ensue:

– The mode switch takes place after a stimulation of the ventricle.

– The mode switch takes place right before the TIMEOUT delay expires (associated with the constraint imposed on the LRL)

The task, newly activated as a consequence of the mode switch, cannot know how much time has passed since the last activation instant. By default, it is necessary to determine when the new impulse must take place. However, no fixed value can ensure that both 22/P and 23/P will be satisfied at the same time in both scenarios above. The use of UPPAAL has allowed us to visualize this problem, which is caused by the variability of the instant where the task can switch its mode.

12.3. Conclusion

In this chapter, we have analyzed the verification of requirements 22/P and 23/P of the pacemaker, having taken as input the previously built AADL models.

To carry out a verification and validation of a system as complex as the pacemaker we need to have a precise methodology clearly defined. We have chosen to present the phases that allow us to translate an AADL model into a UPPAAL model. This example showcases the questions that need to be asked during an model-driven engineering (MDE) process that seeks to translate a model into a formal specification in order to analyze it.

Our endeavor is based on a set of three questions:

– What is the property that must be analyzed? The choice of a particular analysis technique, whether it is a formal one or not, will depend on the nature of the property.

– What is the tool that can support this analysis? The example of the pacemaker shows a great variability in the parameters as well as in the several configurations that need to be processed. This comes from the limitations that are inherent to the tool, but above all from the limitations of the techniques currently used: the general problem being undecidable, we have boiled it down to a family of parameter-related issues that, taken as whole, are indeed decidable.

– How do we translate a model like AADL into a formal analysis? We have shown, in a first phase, that we need to isolate the elements of the input model that are valid for the analysis as well as for the tool. Starting from this list, we have provided the necessary translation elements.

By doing so, we have laid the grounds for an AADL-to-UPPAAL translator, which we have then applied to our verification problem.

Because of the nature of the problem that is based on time intervals, we have opted for a decomposition of the parameters into smaller discrete units, which has allowed us to analyze different configurations. On the one hand, we have shown that the system has indeed verified the requirements for a given mode; on the other hand, we have emphasized the situations where the requirements have not been met.

In the latter case, we must ask: which one is faulty – the model, the requirement, the translation or the tool? We have shown that in this case, there was a structural fault: some information (the date of the last activation) was lost throughout the mode change. We are dealing here with an element that is specific to the AADL language and its definition of a mode change. However, this definition is compatible with a significant number of industrial practices, so it is necessary that the system requirements clarify the notion of "mode", and indicate precisely the elements that are preserved or modified during a mode change, as well as the properties that we deem acceptable to violate during this mode change.

We will let the reader answer this question, a question that has been at the core of many debates, either within the communities of analysis language, or within communities of technical expert system engineers and tool implementers.

It is worth keeping in mind that the verification of a model allows us to eliminate its design flaws, but nothing else.

12.4. Bibliography

[ABD 08] ABDOUL T., CHAMPEAU J., DHAUSSY P., *et al.* "AADL execution semantics transformation for formal verification", *International Conference on Engineering of Complex Computer Systems (ICECCS)*, IEEE Computer Society, pp. 263–268, 2008.

[ALP 87] ALPERN B., SCHNEIDER F.B., "Recognizing safety and liveness", *Distributed Computing*, vol. 2, pp. 117–126, 1987.

[ALU 99] ALUR R., *Timed Automata*, Lecture Notes in Computer Science, vol. 1633, pp. 8–22, 1999.

[BEH 04] BEHRMANN G., DAVID A., LARSEN K. G., "A tutorial on uppaal", in BERNARDO M., CORRADINI F. (eds.), *Formal Methods for the Design of Real-Time Systems, International School on Formal Methods for the Design of Computer, Communication and Software Systems, SFM-RT 2004, Revised Lectures*, Lecture Notes in Computer Science, vol. 3185, Springer, pp. 200–236, 2004.

[BER 09] BERTHOMIEU B., BODEVEIX J.-P., CHAUDET C., *et al.*, "Formal verification of AADL specifications in the topcased environment", *International Conference on Reliable Software Technologies – Ada-Europ*, no. 5570, Springer-Verlag, Brest, France, pp. 207–221, 2009.

[BOS 07] BOSTON SCIENTIFIC, Pacemaker System Specification, January 2007.

[BOZ 10] BOZZANO M., CIMATTI A., KATOEN J.-P., *et al.* "A model checker for AADL", in TOUILI T., COOK B., JACKSON P., (eds.), *Computer Aided Verification*, Lecture Notes in Computer Science, vol. 6174, Springer, Berlin, Heidelberg, pp. 562–565, 2010.

[BRA 83] BRAND D., ZAFIROPULO P., "On communicating finite-state machines", *Journal of the ACM*, vol. 30, no. 2, pp. 323–42, 1983.

[CHK 08] CHKOURI M.Y., ROBERT A., BOZGA M., *et al.* "Translating AADL into BIP – application to the verification of real-time systems", *MoDELS Workshops*, pp. 5–19, 2008.

[KUP 99] KUPFERMAN O., VARDI M.Y., "Model checking of safety properties", HALBWACHS N., PELED D., (eds.), *Computer Aided Verification, 11th International Conference, (CAV)*, Lecture Notes in Computer Science, vol. 1633, Springer, pp. 172–183, 1999.

[RUB 11] RUBINI S., SINGHOFF F., HUGUES J., "Modeling and verification of memory architectures with AADL and REAL", in *Proceedings of the 2011 16th IEEE International Conference on Engineering of Complex Computer Systems (ICECCS '11)*, IEEE Computer Society, Washington, DC, USA, pp. 338–343, 2011.

[SIN 05] SINGHOFF F., LEGRAND J., NANA L., *et al.* "Scheduling and memory requirements analysis with AADL", *Ada Letter*, vol. XXV, pp. 1–10, 2005.

Chapter 13

Model-Based Code Generation

13.1. Introduction

This chapter describes the process of producing real-time embedded software, whose stages are presented in Figure 13.1. The starting point is a model produced in architecture analysis and design language (AADL), which describes a Distributed Real-time Embedded (DRE) application. The large part of this model is written by the user while complying with the language-specific restrictions defined in [ZAL 08]. This is a subset of an AADL language that does not include non-deterministic or too complex constructs for DRE systems. The definition of this subset is based on the Ravenscar profile for the development of critical systems [BUR 04].

To specify the non-architectural properties of the components described, the developer uses the standard properties of the AADL language. In case the standard properties that specify a particular characteristic (task priority, data type category, etc.) are absent, the developer uses a set of properties that completes the standard sets (also defined in [ZAL 08]).

These last properties as well as other missing ones (priority, etc.) have been integrated in the very recent version of the standard AADLv2. All these properties enrich the expressiveness of the model and allow us to analyze various characteristics of the components:

– functional characteristics such as the names of the implementations of AADL subprograms given as source files or libraries;

– non-functional characteristics such as threads periods or their deadlines;

Chapter written by Laurent PAUTET and Bechir ZALILA.

– architectural characteristics such as the processor binding of a process.

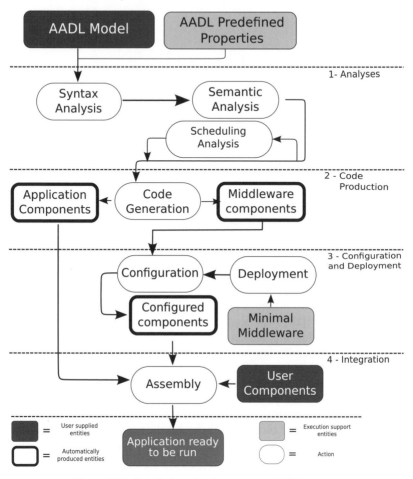

Figure 13.1. *Detailed production process of DRE systems*

As shown in Figure 13.1, the production process that we proposed has four main steps.

– *Analyses*. Once the DRE application has been modeled in AADL, its model is subjected to a series of analyses meant to ensure that the model is coherent. These analyses consist of verifying the syntax and the semantic of the model by confronting it with the standard AADL. These stages are described in Chapters 10 and 11. The analyses described in Chapter 12 are then applied to the models.

We also verify that the analyzed model is limited to the subset of the AADL language. This enables us to have a code generated in accordance with the Ravenscar profile [BUR 04]. Optional analyses can be also carried out by external tools (this is the case of scheduling analysis). These analyses performed on the AADL model reinforce application reliability.

– *Code production.* After the analysis step, the AADL model is used to automatically produce two sets of components:

- The application software components that generally gather the data types and the subprograms used by the user. These components are produced automatically from the AADL model, more specifically from the `data` and `subprogram` components. Being able to produce them automatically, with precise naming constraints that are known by the user, has a double advantage. First, it allows us to trace the code throughout the entire development process. It is, indeed, easy to find the source code entities that have been generated from an AADL entity: they generally have the same name or very similar names. This traceability proves to be necessary in the case of the certification process of several families of critical embedded types of software. Second, this allows us to use the automatically generated entities in other places of the generated code written by the user. The conversion rules we use are similar to the mapping rules from RDL to a specific programming language in CORBA.

- The highly customizable middleware components, which can be customized according to the properties of the application. These components are produced automatically, depending on the topology of the application and, in particular, the interaction between the `process`- and the `thread`–type components. Depending on the platform, their optimization is carried out by analyzing the hardware components that can be found in the model (`processor`, `bus`, etc.).

– *Configuration and deployment.* This stage consists of selecting these middleware services (also known as the AADL execution platform) that are used by a given node. The selection is done automatically due to the deployment information extracted in particular from the hardware components of the AADL model (characteristics of the `processor` and `bus` components). These selected services, as well as those generated automatically, are configured according to the properties of the node. At the beginning of these two stages, a set of middleware services is being produced, a set that is optimized and configured according to the needs of each of the application nodes. A service is an architectural element of the middleware layer, which performs one step of the communication between two nodes by calling the corresponding routines of the operating system. We find in this category, for instance, the protocol that allows us to manage a communication according to the given protocol (IIOP, etc.) and the execution service that allows us to run on the server the activities required by the clients. Generally, the parameters of the services can be fine-tuned according to the needs of an application. Certain services cannot even be used. For this reason, they are isolated so they can be produced and configured independently from one another.

– *Integration*. This step consists of gathering the software application components that have been produced during the code generation step, the middleware components that have been configured throughout the deployment and configuration steps, and the components supplied by the user. This linking process allows us to produce the executable files corresponding to the application nodes. The number of executable files may be more than one in the case of distributed systems, particularly in the case of the pacemaker. At the end of this step, we will have an application that is ready to run.

In the remainder of this chapter, we will detail stages 2–4. In sections 13.2 and 13.3, we discuss the conversion rules of the AADL models in order to produce the software as well as middleware components. Section 13.4 describes the deployment and configuration processes of the middleware components. In section 13.5, we present the way in which all of the software components are automatically integrated in order to form distributed application software that is ready to be run. Finally, section 13.6 concludes the chapter.

13.2. Software component generation

In this section, we discuss the rules for converting the several AADL components into blocks of an imperative programming language. Our conversion rules are general although the examples that illustrate them are written in Ada. Each of these conversion rules is followed by several examples of AADL entities and respectively their converted entities.

It is worth noting that sometimes the differences between the application languages generated are not very significant. For instance, if we wish to define a 32-bit integer type in Ada language, the following code will be produced:

```
type <Type_Name> is new Interfaces.Integer_32;
```

Whereas for the C language, we will have the following build:

```
typedef int_32 <Type_Name>;
```

In other cases, the Ada code is fundamentally different from the C code. For example, if we want to create an execution thread that sends a Hello call to a subprogram every second, in Ada we will have the following code:

```
task <Nom_Du_Thread> is
   pragma Priority (...);
end <Nom_Du_Thread>;
```
4
```
// Creation of an execution thread
task body <Thread_Name> is
   Period           : constant Ada.Real_Time.Time_Span := ...;
```

```
          Next_Activation  :  Ada.Real_Time.Time;
 9   begin
          Next_Activation  := Ada.Real_Time.Clock + Period;
          loop
             Hello;
             delay until Next_Activation;
14           Next_Activation := Next_Activation + Period
          end loop;
      end <Thread_Name>;
```

On the other hand, the C language does not intrinsically support tasks. We must encapsulate the periodic behavior in a subprogram and go through calls to system libraries such as POSIX in order to create a task that executes the subprogram. We see, in the example below, that the absence of this intrinsic support makes the C code much more complicated than the Ada code.

```
      pthread_mutex_t  mutex;
      pthread_cond_t   cond;

 4   void delay_until (struct timespec *next_activation) {
          pthread_mutex_lock (mutex);
          pthread_cond_timedwait (&cond, &mutex, &next_activation);
          pthread_mutex_unlock (&mutex);
      }
 9
      void wrapper (void) {
          struct timespec period;
          struct timeval next_activation;

14        pthread_mutex_init (&mutex, NULL);
          pthread_cond_init (&cond, NULL);
          period.tv_sec  = ...;
          period.tv_nsec = ...;
          gettimeofday (&next_activation);
19        next_activation.tv_sec = next_activation.tv_sec + period.ts;
          next_activation.tv_nsec = next_activation.tv_usec * 1000 + period.tv_nsec;

          while (1) {
             hello ();
24           delay_until (&next_activation);
             next_activation.tv_sec  = next_activation.tv_sec  + period.ts;
             next_activation.tv_nsec = next_activation.tv_nsec + period.tv_nsec;
          }
      }
29
      /* Creation of the thread */
      pthread_create (&tid, NULL, wrapper, NULL);
```

13.2.1. *Data conversion*

The data types are deduced from instances of data type AADL components. These components are instantiated in several locations in the model:

– in the data ports or event data ports or parameters found either in processes, in threads or even in subprograms;

– in the parameters of subprograms;

– during the declaration of shared variables between several entities of the model.

These declarations are transformed into type definitions. The defined types are used for defining the different instances. The asynchronous character imposed on the communications forces us to define a "default" value for each of these defined data types. This value is used when the data are not yet available (in the initialization of the application, for example, or during a partial breakdown that makes the communication temporarily unavailable).

Thus, a declaration of this type of constant that has a "default" value for this type is produced for each data component declaration. This mechanism is only useful for the periodic tasks, which are the only processes that can be triggered by a time event without having valid received data.

By convention, the name for this type is the same as the name of the AADL component, except for the case where the latter does not observe the naming rules with the language; in this case, the mapping rules must provide the means to convert the names of the components into names that are compatible with the programming language (somewhat similar to the mapping rules of the CORBA IDL into the different programming languages). For instance, the Ada language forbids the definition of the identifiers that contain successive occurrences of the "_" character. The identifiers of the AADL model of the user, if disregarding this restriction, are modified by taking a character that is different from "_" and inserting it between successive occurrences of the latter.

```
  data Time
  properties
    Data_Model :: Data_Representation => Integer ;
4 end Time ;

  data Array_Of_Time
  properties
    Data_Model :: Data_Representation => Array ;
9   Data_Model :: Dimension => (10) ;
    Data_Model :: Base_Type => (data Time) ;
  end Array_Of_Time ;
```

Listing 13.1. *AADL data types*

```
  ——   Time AADL component transformation
  type Time is new Float ;

4 ——   Array_Of_Time AADL transformation
  type Array_Of_Time is array (1 .. 10) of Time ;
```

Listing 13.2. *Generated Ada code*

Listing 13.1 gives an example in AADL of an integral type of data as well as of an array type that contains 10 integers. Listing 13.2 describes the conversion of these two components into Ada language. In the absence of additional information from the user (size of the integer type, interval, etc.), the type is simply an extension of the standard Integer type. The user can be more specific about the definition of a type by using the properties provided in the predefined Data_Model property set. We have defined this set of properties in order to solve the lack of analysis in the first version of the AADL standard. This set will be part of the standard property sets of the next version AADLv2.

The nature of the type defined depends largely on the value of the property Data_Model::Data_Representation. If the AADL component represents a data structure, it will be converted into a recording type with data files whose types correspond to projections of their respective declarations.

In case there are any accessing subprograms, their implementation must be in accordance with the concurrency management protocol of the components. We want the data access to be protected against the concurrence and for the priority switch due to this protection to be limited. We also want to guarantee the absence of deadlock, which might be caused by this protection. One solution would be to use a priority inheritance protocol. The Ravenscar profile [BUR 04] recommends the use of priority ceiling protocol (PCP) [SHA 90]. Not only does it guarantee the absence of deadlock but it also reduces the blocking time of a task (caused by successive blockings).

A few imperative programming languages (such as Ada) intrinsically dispose of the necessary constructs for these protected variables. For the other programming languages (such as C), we will use system libraries in order to create and initialize the software mutexes needed to protect the data. This makes the C code much more complicated than the Ada code especially in terms of verification and certification.

```
   data Time extends
   properties
     Data_Model::Data_Representation => Integer;
   end Time;
5
     — A data record with two integers
   data Record_Type
   properties
     Data_Model::Data_Representation => Struct;
10 end Record_Type;

   data implementation Record_Type.Impl
   subcomponents
     X : data Time;
15   Y : data Time;
   end Record_Type.Impl;
```

Listing 13.3. *AADL type samples*

```
        —   Default value for Component_Type.
     Time_Default_Value : constant Time := 0.0;

4    —   Default value for Record_Type_Impl.
     Record_Type_Impl_Default_Value : constant Record_Type_Impl :=
            (X => Time_Default_Value,
             Y => Time_Default_Value);
```

Listing 13.4. *Generated Ada code*

Listing 13.3 gives an example of a record type modeled in AADL. This is a three-integer structure (each representing a coordinate in space). The default constant generated for this structure is a structure having in each data field the default value associated with its type. Listing 13.4 shows the translation of such a constant in Ada.

13.2.2. *Conversion of subprograms*

The AADL subprograms are entities destined that host the code of the user. The FEATURES clause might contain parameters for exchanging data with the exterior. In AADL, the subprograms are passive entities that must be called by the tasks to be executed. Thus, the subprogram entities are naturally converted into subprograms of the imperative programming language chosen for the code generation.

The behavior of the subprogram produced depends on the nature of the AADL component. We classify the AADL subprograms into three categories depending on the values of the Source_Language standard properties and the presence or absence of the CALLS clause in the component implementation. These categories are opaque subprograms implemented in a particular programming language, pure call sequence subprograms and empty subprograms.

Because of the presence of out-ports, the AADL subprograms (of all categories) can trigger events and data-events on the caller's side. This allows us to control the sending of events from the user code by using the methods defined by the AADL standard in order to trigger the said events. The conversion of such subprograms must take into account the fact that the same subprogram may be invoked by several tasks on the same node. In other words, this conversion must guarantee that the data are not corrupted during concurrent accesses.

We have chosen to use an additional "opaque" parameter that independently represents the state of the ports of a subprogram to each call. This guarantees the coherence of the data in the case of coherent access. We will detail this category of subprograms during task conversion. The generated code in both cases is very similar.

The code generator deduces the category of a subprogram from its characteristics. Then, depending on the found category, the generator produces the corresponding code. In what follows, we will explain the generator behavior for each category of subprograms.

13.2.2.1. *Conversion of opaque subprograms*

In this section, we discuss the simplest implementation of a subprogram in AADL. The entire behavioral part is supplied by the user and the AADL model contains only the definition of the subprogram's interface. For these subprograms, three properties are defined – Source_Language, Source_Name and Source_Text – and the CALLS clause is absent. The AADL model indicates the programming language, the name and the source file for its implementation. These components facilitate the generation of a skeleton subprogram that does a call to the indicated implementation. In the case of compilators such as Ada GNAT, where the name of a file is deduced from the name of the compilation unit it contains, the informing of the Source_Text field is optional.

```
  subprogram Battery_Test
  features
3   Level : out parameter Base_Types::Float_32;

  properties
    source_language => Ada95;
    source_name     => "Pacemaker.Do_Battery_Test";
8  end Battery_Test;
```

Listing 13.5. *Opaque subprogram*

```
  with Pacemake;
2 procedure Battery_Test (Level : out Base_Types.Float_32)
    renames Pacemaker.Do_Battery_Test;
```

Listing 13.6. *Generated Ada code*

Listing 13.5 shows the AADL model of an opaque subprogram implemented in Ada. This subprogram will be transformed in a code package that calls the Pacemaker.Do_Battery_Test implementation provided by the user as indicated in Listing 13.6.

13.2.2.2. *Conversion of the pure call sequence subprograms*

In this case, the subprogram contains one single call sequence in its CALLS clause. None of the three properties mentioned above need to be defined for this type of subprogram. The behavior of the subprogram is, therefore, defined entirely by its AADL model: a sequence of calls to other subprograms. This type of component facilitates the generation of a subprogram that carries out the call sequence towards the respective converted entities corresponding to the called components. Of course, it is possible to connect the parameters at the entrance and exit of the called subprograms. The generated code must provide all of the intermediary variables needed for ensuring the good data flow analyzed by these connections.

```
   subprogram Battery_Test
2  features
      Level: out parameter Base_Types::Float_32;
      properties — ...
      end Battery_Test;

7  subprogram Internal_Transmitter
   features
      Level   : in   parameter Base_Types::Float_32;
      Output  : out parameter Base_Types::Float_32;
      properties — ...
12 end Internal_transmitter;

   subprogram Sender
   features
      Level : out parameter Base_Types::Float_32;
17 end Sender;
   subprogram implementation Sender.impl calls {
      Test      : subprogram Internal_Sender;
      Transmit : subprogram Internal_transmitter;};
   connections
22    C_1 : parameter Test.Level —> Transmit.Input;
      C_2 : parameter Transmit.Output —> Level;
      end Sender.impl;
```

Listing 13.7. *Pure call sequence subprogram*

```
1  procedure Battery_Test (Level : out Float) is
      —  ...
      end Battery_Test;

   procedure Internal_Transmitter
6    (Level :      Float;
      Output : out Float)
   is
      —  ...
      end Internal_Transmitter;
11
   procedure Sender_Impl (Output : out Float) is
      C_1_Var : Float; — Corresponds to the C_1 connection.
   begin
      Internal_Sender (C_1_Var);
16    Internal_Transmitter (C_1_Var, Output);
      end Sender_Impl;
```

Listing 13.8. *Code Ada gnr*

Listing 13.7 presents a Sender.Impl AADL subprogram whose behavior consists of calling the Internal_Sender subprogram and then the Internal_Transmitter subprogram. We will not discuss here the process of how these last two subprograms are implemented. Listing 13.8 corresponds to the Ada code generated from this model. The Sender.Impl subprogram facilitates another subprogram called Sender_Impl.

This one calls both the subprograms produced on the basis of `Internal_Sender` and `Internal_Transmitter`.

The code generator automatically processes the generation of intermediary variables that correspond to the connections that link the subprograms called. In this case, an intermediary variable `C_1_Var` is generated: It corresponds to the connection between the `C_1` that links the `Output` parameter to the `Internal_Sender` and the `Input` parameter of the `Internal_Transmitter`.

13.2.2.3. *Conversion of the empty subprograms*

The empty subprograms are subprograms for which the user has not specified any behavioral characteristics. They lead to the production of a subprogram that raises a fatal error at runtime. This kind of subprogram seems of little use upon first sight. However, we will use them to test the models during the first stages of development. Then, we will "fill" in these subprograms by adding their respective behaviors.

13.2.3. *Conversion of execution threads*

The `thread` components are converted into language constructs that enable us to build a task. A few programming languages offer intrinsic builds for creating such execution threads (such as the Ada language via the intermediary of the task mechanism). Other languages make calls toward system libraries in order to create these execution threads (like the POSIX [IEE 94], which is used by several languages).

As previously mentioned, only three types of tasks are accepted in order to guarantee a static scheduling analysis of the application (periodic and sporadic). For each of these three types, a design pattern is built in the chosen imperative language. It describes the behavior of the task according to its category but also according to the semantic that the AADL standard has associated with the `thread` components [SAE 04, section 5.3]. This pattern must be instantiated for each task of the application depending on the characteristics and the properties of the corresponding `thread` component. The needed parameters for the instantiation of this pattern are:

1) the list of interface elements of the component; this list is required by the sporadic threads;

2) the identifier of the *thread*;

3) the period for the periodic tasks or the minimal interarrival time for the sporadic tasks;

4) the deadline;

5) the priority;

6) the storage size;

7) for the sporadic tasks, a subprogram that executes the waiting for an external event; a call to this subprogram blocks the task until the arrival of an event and returns the interface element that has received it;

8) the subprogram that carries out the cyclical work;

9) a possible subprogram that initializes the *thread*.

Certain parameters are deduced directly from the AADL model by simply reading the properties (period, timeout, priority[1], its storage size and the subprograms that carry out the initialization). The other parameters must be generated automatically by analyzing the *thread* component. In what follows, for each of the constituents of a thread component, we will give the conversion roles in an imperative programming language.

13.2.3.1. *Conversion of the interface elements*

The list of interface elements of a thread component is converted into an enumeration. We have chosen this conversion in order to be able to, on the one hand, create tables indexed by the different ports of a thread component and, on the other hand, use the enumerators in conditional structures with multiple choices (*switch case*). This allows an execution time that is constant regardless of the number of ports in a thread component.

```
   thread Rate_Computation
   features
3     Activity : in data port acceleration_intensity ;
      Regulated_Rate_Delay : out data port PACEMAKER_Interfaces : : pacing_delay ;
   properties
      Dispatch_Protocol => Periodic ;
      Priority => 2;
8     Period => 100 ms;
   end Rate_computation ;
```
Listing 13.9. *AADL thread declaration*

Listing 13.11 shows an AADL model of a sporadic task that has a set of in-ports and out-ports. This set of ports is translated, in Ada, for example, through the definition of an enumerated type, as shown by listing 13.12.

---

1 To schedule a set of tasks, their priorities must be deduced from their respective periods. In the first step, we suppose that the user gives periods coherent with Rate Monotonic Scheduling (RMS) in its AADL model and we allocate a scheduling of *First-In First-Out (FIFO) within priorities* type.

```
1   type Rate_Computation_Port_Type is
        ( Activity , Regulated_Rate_Delay );
```

Listing 13.10. *Ada code for the thread port*

Let us note that this same conversion rule is applied to the subprograms that have ports. It leads to an enumeration that will be used by the event sending routine.

Conversion of the behavioral part. The construction of a subprogram that carries out the cyclical work of a *thread* strongly depends on the behavior specified by the user. Indeed, we authorize three different ways of analyzing the behavior of a thread component.

Tasks with one call sequence. The behavior of these tasks is quite simple and very similar to that of the subprograms with a pure call sequence. With each occurrence, they execute the list of subprograms called in their call sequence. This way of implementing a thread component is well adapted to periodic tasks. The entire behavior is automatically managed by the code generator and the user does not have to manipulate the ports of the task in order to send or receive information.

The subprogram generated is similar to an AADL subprogram that calls other subprograms (given in Listings 13.7 and 13.8). The code generator produces the intermediary variables needed for the connection between the subprograms called. It produces the collection of data received in the in-ports at the entry of the thread component and the calls of the subprograms in the right order. Finally, the generator sends potential data or events toward the out-ports of the component.

Tasks with one entry point per port. In this case, each of the event in-ports (or data events) of the thread component finds itself affiliated to a subprogram by the intermediary of the Compute_Entrypoint property applied to each port. The thread component described in Listing 13.11 is part of this category. Thus, with each triggering of the task, the entry point corresponding to the port that has received the triggering order is executed. This way of implementation is adapted to sporadic and hybrid tasks.

The generator produces a subprogram that has a parameter denoting the port that has triggered the task. Then, in the subprogram produced, a test is carried out on the value of this port and the corresponding call point is called.

Listing 13.11 shows the A_Thread and adds to it the Dispatch_Protocol property for specifying that the task is sporadic, and it associates an entry point with each entry of the port. Since this component has one entry point per port, the behavior of the generated subprogram executes the entry point corresponding to the received event. The use of enumeration for representing the ports allows us to arrive at the behavior desired in constant time. As shown in Listing 13.12, each of these entry points for the event-data type of port has a parameter in its signature. This allows the code of the user to find the value of the data received by the port. The reading of the data is carried out with the help of the Get_Value routine.

```
    thread A_Thread
    features
3      Input_1   : in event data port Integer {Compute_Entrypoint =>
             "Pkg.On_Input_1";};
       Input_2   : in data port Integer;
       Input_3   : in event port {Compute_Entrypoint => "Pkg.On_Input_3";};
       Output_1 : out event data port Integer;
       Output_2 : out data port Integer;
8      Output_3 : out event port;
    properties
       Dispatch_Protocol => Sporadic;
       —   ...
    end A_Thread;
```

Listing 13.11. *AADL thread implementation*

```
    with Pkg;
    procedure A_Thread_Job (Port : A_Thread_Port_Type) is
3   begin
       case Port is
          when Input_1 => Pkg.On_Input_1 (Get_Value (Input_1));
          when Input_3 => Pkg.On_Input_2;
          when others  => raise Program_Error;
8      end case;
    end A_Thread_Job;
```

Listing 13.12. *Ada job of the thread*

In the case of tasks with only one entry point, the latter is analyzed by the intermediary of the standard property Compute_Entrypoint associated with the thread component itself. We need to enable the user to manage for themselves the entire behavior during an occurrence (port queue management, dialog with the middleware routines for sending information, etc.).

This design method is used in the rare cases where the user wants to carry out advanced behaviors on the queues of the interface elements (such as the extraction of more than one element, or the reading of a queue different from the one of the port that has just caused the trigger). In this case, the user hides part of the behavior from the verification step.

Conversion of operational modes. The operating modes of a `thread` component, insofar as they are present, can only control the behavior of the corresponding task. Thus, they do not compromise or complicate the application's static analyzability. In this case, somewhat similar to the way in which we have converted the interface elements, the modes are transformed into an enumeration. The state machine that runs the mode change is transformed into a state machine in the imperative programming language. It is placed at the beginning of the subprogram generated in order to carry out the cyclic behavior. Thus, at the beginning of its execution, the value of the mode is calculated depending on the events received and the current value. Then, the assessed subprogram carries out the processing according to the found mode.

```
 1   thread A_Thread
     features
        Go_Lazy    : in event port ;
        Go_Normal  : in event port ;
        Go_Crazy   : in event port ;
 6   properties Dispatch_Protocol => Sporadic ;
     end A_Thread ;
     thread implementation A_Thread . Impl
     modes
        Normal : initial mode ;
11      Crazy  : mode ;
        Lazy   : mode ;
        Normal , Lazy   -[Go_Crazy ]-> Crazy ;
        Normal , Crazy  -[Go_Lazy  ]-> Lazy ;
        Crazy ,  Lazy   -[Go_Normal]-> Normal ;
16   properties
        Compute_Entrypoint => "Pkg.Normal_Handler" in modes (Normal) ;
        Compute_Entrypoint => "Pkg.Crazy_Handler" in modes (Crazy) ;
        Compute_Entrypoint => "Pkg.Lazy_Handler" in modes (Lazy) ;
     end A_Thread . Impl ;
```

Listing 13.13. *Modes in the thread AADL*

Listing 13.13 shows an AADL model of a sporadic `thread` component having three operating modes. The component has a single entry point. The mode change is described by the state machine given in the MODES clause of the `A_Thread.Impl` component. Depending on the value of the current mode, the task executes one of the subprograms analyzed in the PROPERTIES clause of the same component.

The Ada code produced for managing the mode change is presented in Listing 13.14. We see the enumeration representing the modes (`Mode_Type`). The global variable representing the current mode is initialized at `Normal` as specified in the AADL model. Besides, the generated code implants the state machine that runs the mode change and carries out the desired behavior. There too, the use of an enumeration for representing the operating modes allows us to do the computing and the transitions in constant time.

```
    type Mode_Type is (Normal, Crazy, Lazy);
    Mode : Mode_Type := Normal;
    procedure A_Thread_Job (Port : A_Thread_Port_Type) is
    begin
5     case Port is
        when Work_Normally =>
          case Mode is
            when Crazy | Lazy => Mode := Normal;
            when others        => null;
10        end case;
        when —  ...
        when others => null;
      end case;
      case Mode is
15      when Normal => Pkg.Normal_Handler (Port);
        when Crazy  => Pkg.Crazy_Handler (Port);
        when Lazy   => Pkg.Lazy_Handler (Port);
      end case;
    end A_Thread_Job;
```

Listing 13.14. *Thread "job" in Ada*

Instantiation of the task. An AADL `thread` component is converted into:

– the Ada instance of the design pattern for creating an execution thread or;

– methods that allow the read and/or write access to the interface of the component in order to send and receive the information.

```
1   package A_Thread_Task is new Sporadic_Task
      (Port_Type                 => Receiver_Port_Type,
       Entity                    => SC_2_Receiver_ID,
       Task_Period               => Ada.Real_Time.Milliseconds (20),
       Task_Deadline             => Ada.Real_Time.Milliseconds (20),
6      Task_Priority             => System.Default_Priority,
       Task_Stack_Size           => 64_000,
       Job                       => A_Thread_Job,
       Wait_For_Incoming_Events  => Wait_For_Incoming_Events);
```

Listing 13.15. *Instance of an Ada sporadic thread*

Listing 13.15 shows the Ada instance generated for `A_Thread.Impl` in Ada. The code generator allocates values by default to the optional parameters that are not provided by the user. Thus, the deadline of the task takes on a value that is equal to that of the period, and its priority is set to the default priority of the system.

The `Wait_For_Incoming_Events` subprogram that is the last parameter of the instantiation of `A_Thread.Impl` is part of the methods of the interaction service that is produced automatically. This service constitutes a layer over the interaction service of the minimal middleware. It provides the following routes that facilitate the read and write access to the interface elements of a `thread` component.

The names of these routines as well as their behaviors are given by the AADL standard:

– `Send_Output` explicitly sends the events and the data throughout the network in case of distant communication. If the communication is local, this method makes the direct copy toward the destination. It takes as parameter a reference toward the port that must be processed.

– `Put_Value` marks the fact that a port has a piece of information (data or event) that is ready to be sent. The sending of the data will be carried out by a call to `Send_Output`. It takes as its parameters a reference towards the port in question as well as the value of the data that needs to be sent if there is a case of a data port or an event-data port.

– `Get_Count` returns the number of messages that are not yet consumed in the queue of an event port or an event-data port.

– `Get_Value` sends the value received, which can be found at the top of the queue of an event or event-data port or simply the value of a data port. It does not cause any consumption: using this method several times gives the same result.

– `Next_Value` consumes the message at the top of the queue of an event or of an event-data port;

– `Wait_For_Incoming_Events` regards sporadic tasks. It waits until an event arrives and returns the port where the event has arrived.

The behavior being generic, it is not very customizable; depending on the properties of an application, they are grouped together as an archetype that is part of the basic middleware. For each task, an Ada instance of these routines is created. We create an Ada instance for each AADL component before being able to statically parameterize it.

The Ada instance is parameterized by the following entities:

1) An enumerated type that lists the ports of the `thread` component.

2) A discriminated record (such as the unions in C), whose discrimination is of the enumerated type mentioned above. For each data or event-data port, this record will contain an additional data field that has the data type. For the ports of a pure event type, this record is void;

3) An identifier for the *thread*.

4) A table indicating the category and the propagation direction for each port (*in data port*, *out data port*, *in event port*, etc.).

5) A table indicating, for each event in-port, the size of the queue (analyzed with the help of the property AADL Queue_Size).

6) A table indicating, for each out-port, the list of its destinations.

7) A *Marshall* routine that inserts the information associated with a port into a queue.

8) A *Next_Deadline* routine that returns the date of the next deadline of the *thread*.

```
1    type Rate_Computation_Interface (Port : Rate_Computation_Port_Type) is
        record
           case Port is
              when Activity             => Activity_DATA : Integer;
              when Regulated_Rate_Delay => Regulated_Rate_Delay_DATA : Integer;
6          end case;
        end record;

     type Rate_Computation_Integer_Array is
        array (Rate_Computation_Port_Type) of Integer;
11   type Rate_Computation_Port_Kind_Array is
        array (Rate_Computation_Port_Type) of Port_Kind;
     —   Types used for the constants below

     Rate_Computation_Port_Kinds : constant Rate_Computation_Port_Kind_Array :=
16      (Activity             => In_Event_Data_Port,
         Regulated_Rate_Delay => Out_Event_Data_Port);

     Rate_Computation_FIFO_Sizes : constant Rate_Computation_Integer_Array :=
        (Activity             => 16,   — Default value
21       Regulated_Rate_Delay => −1);  — Convention (out part)

     type Rate_Computation_Address_Array is
        array (Rate_Computation_Port_Type) of System.Address;

26   Rate_Computation_N_Destinations : constant Rate_Computation_Integer_Array :=
        (Activity             => −1, — Convention (in part)
         Regulated_Rate_Delay => <Nbr Destinations >); — Deduced from
                application topology

     Rate_Computation_Destinations : constant Rate_Computation_Address_Array :=
31      (Activity             => System.Null_Address,
         Regulated_Rate_Delay => <Tableau Des Ports Destination >'Address);

     —   Generic package instance

36   package Rate_Computation_Interrogators is new Thread_Interrogators
        (Port_Type            => Rate_Computation_Port_Type,
         Integer_Array        => Rate_Computation_Integer_Array,
         Port_Kind_Array      => Rate_Computation_Port_Kind_Array,
         Address_Array        => Rate_Computation_Address_Array,
41       Thread_Interface_Type => Rate_Computation_Interface,
         Current_Entity       => Rate_Computation_Impl_ID,
         Thread_Port_Kinds    => Rate_Computation_Port_Kinds,
         Thread_Fifo_Sizes    => Rate_Computation_FIFO_Sizes,
         N_Destinations       => Rate_Computation_N_Destinations,
46       Destinations         => Rate_Computation_Destinations,
         Marshall             => Marshall,
         Next_Deadline        => Rate_Computation_Task.Next_Deadline);
```

Listing 13.16. *Ada instance of "Interrogators" for* ***A_Thread_Impl***

Listing 13.16 shows how these interrogators are instantiated in the case of Ada. Additional entities must be provided with the package instance in order to complete the instantiation such as data types. The new instantiated pack `A_Thread_Interrogators` provides the subprograms mentioned above.

For the AADL subprograms that have ports, a set of interrogators is created in a similar manner.

13.2.4. *Conversion of the instances of shared data*

The instances of protected data are the subcomponents of `data` types that are declared in the implementations of the `process` components. These are the data shared between the tasks of the same process.

A `data` subcomponent is characterized by its name and the type of component it instantiates. It is converted in a declaration of a shared variable (protected or not, according to the specification of the user). A set of methods for accessing this variable is then generated. These methods are used by the tasks or the subprograms sharing the access to this variable.

13.3. Middleware components generation

As previously mentioned, part of the middleware is produced automatically. These are highly customizable middleware services. These components are dedicated to the characteristics of the current node. The components are the following:

– The Ada instances of *threads*: for each AADL component of a `thread` type, an instance of a generic Ada package that corresponds to its category is being created. All of these instances are created in the **PolyORB_HI_Generated.Activity** package.

– The advanced marshalling and unmarshalling routines corresponding to the types of data present in the AADL model. These are the subprograms that use instances of the generic Ada package **PolyORB_HI.Marshallers_G** that belongs to the minimal middleware. This package offers the elementary representation mechanisms. The subprograms generated implement the advanced representation service. They belong to the **PolyORB_HI_Generated.Marshallers** package.

– The static naming tables correspond to each of the supported transport services. These tables are produced in the **PolyORB_HI_Generated.Naming** package.

– The high-level transport service whose routines allow us to correctly select the low-level transport service depending on the source and the destination of a message. This is produced in the **PolyORB_HI_Generated.Transport** package.

13.4. Configuration and deployment of middleware components

In this section, we explain how the basic middleware components that the application needs are selected (deployment). We will thus explain how these components, as well as those produced automatically, are parameterized depending on the characteristics of the application (configuration).

13.4.1. *Deployment*

The selection of the basic middleware components is done while going through the instance tree of the AADL model of the application. In particular, the properties of the hardware components are analyzed in order to determine precisely which of the basic middleware components they need. We use the set of deployment properties that we have defined in [ZAL 08].

Bus. The properties of an AADL bus allow us to select the necessary versions of the basic middleware services:

– The low-level transport service: indeed, among the properties related to deployment that we have added, a `Transport_API` property allows us to detail the low-level transport mechanism that uses the bus (Ethernet, SPACEWIRE).

– The protocol: the bus also contains information on the protocol used. The `Protocol` property that we have introduced details the protocol used for exchanging data via the buses.

– The elementary representation: generally, each used protocol is associated with a representation service that indicates the way in which the data will be exchanged. For example, the default protocol that consists of sending the data in the network via a single copy does not need an advanced representation service. It suffices to convert the data into byte streams without modifying it. On the other hand, for a protocol such as GIOP, we use the CDR representation specified by CORBA.

Processors. The processors specify the platform where the node is being executed. This allows us to refine the service selection. For example, if the same service is implemented differently for two different platforms, the nature of the processor specifies the version that needs to be chosen. The platform of a processor is specified by means of the `Execution_Platform` property that we have previously introduced.

Configuration. The configuration of the middleware services is done statically and very simply due to the different enumerations and constants that we produce by transforming the AADL entities. In the following, we give these enumerations and constants and we specify which services they configure.

Nodes. On each node of the application, we produce an enumeration that contains the list of nodes to which it is connected. We thus produce a constant that has the same enumeration type and that allows the node to identify itself.

This enumeration is used to configure the low-level transport service by opening communication channels only for the nodes that are really connected.

Threads. On each node of the application, we produce an enumeration that contains the list of threads components of that node as well as the threads components of other nodes that communicate with the current node. We thus produce a table that associates the threads components with the nodes they belong to.

This enumeration as well as this table are used to complete the initialization of the low-level transport services by opening the correct number of communication channels toward each remote node.

Size of the messages. On each node, we analyze all of the data being manipulated and we calculate the maximum size of a sent or received message. Then, we generate a constant that represents this size. This constant facilitates the initialization of the protocol services and of the elementary representation by statically allocating the queues.

Let us note that the assessment of this constant is not uniquely based on the size of the data sent. It also takes into account the size of the headers of the messages.

13.5. Integration of the compilation chain

After the generation of software and middleware components of the distributed application, the last step consists of integrating this code to the already existing components of the minimal middleware as well as to those of the user. Then, the code is compiled in order to have a number of executable files (equal to the number of nodes in the application) ready to be downloaded and executed in their respective locations.

Making this step automatic seems simple upon first glance. However, a more in-depth analysis of the problem shows that it requires a significant amount of work compared to other steps of the process (analyses, code generation, etc):

– Certain programming language compilers require accompanying files in addition to the source files in order to compile them (this is the case of the Ada GNAT compiler and the project files or the Makefiles).

– The integration of the basic middleware components requires that the right compiler be selected according to the chosen programming language and that the crossed version of the most convenient cross-compiler be selected according to the architecture of the node in the application model.

– The integration of precompiled components and/or components produced by third-party tools requires the addition of options during the service invocation of the compiler.

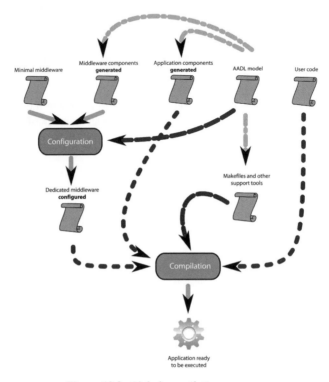

Figure 13.2. *Global compilation process*

Figure 13.2 describes the global compilation process. Having helped generate the software and middleware components, the AADL model is also used to drive the configuration step of the middleware. This step consists of selecting the components of the basic middleware that the application needs and integrating them to the automatically generated components in order to produce a middleware program that is especially configured for the application. This is particularly carried out due to the incorporation of the description of the hardware components in the model.

The AADL model also facilitates the generation of support files (Makefiles, project files, etc.). These files run the compilation step that takes as input the user code (including the code produced by third-party tools) as well as the code of the middleware configured.

To produce these files, an analysis of the entire application model is necessary; this analysis will reveal the relations between the names of the source files, how they depend on each other, etc. The hardware part of the AADL model is also analyzed in order to deduce the compiler that will be used (a native compiler or a cross-compiler). At the end of the compilation step, we create a distributed application that is ready to be run.

13.6. Conclusion

In this chapter, we have presented the different stages of the automatic process of producing critical DRE systems from AADL models. Starting from an AADL model written according to the restrictions studied in the previous chapter, we have presented the different analyses that need to be carried out on this model to ensure the good functioning of the system. We have shown three kinds of analyses that we may carry out on an AADL model. They vary from simple semantic analyses to more advanced scheduling analyses.

The most important part of this process is the automatic code generation that, starting from the AADL application model, produces the different software and middleware components that will be integrated with the already existing components, in order to form the different nodes of the application software. We have presented the conversion rules of the different AADL components into an imperative generic programming language. We have then shown how the middleware components are deployed and configured depending on the characteristics of the running DRE application. Our code generation process produces a readable source code that is completely adapted to the needs of the application.

13.7. Bibliography

[BUR 04] BURNS A., DOBBING B., VARDANEGA T., "Guide for the use of the ada ravenscar profile in high integrity systems", *Ada Letters*, vol. 24, no. 2, pp. 1–74, 2004.

[IEE 94] IEEE I., IEEE Standard for Information Technology – Portable Operating System Interface (POSIX): System Application Program Interface (API), Amendment 1: Realtime Extension (C Language), IEEE Std 1003.1b-1993, IEEE Standards Office, New York, 1994.

[SAE 04] SAE, Architecture analysis & design language (AS5506), September 2004. Available at www.sae.org.

[SHA 90] SHA L., RAJKUMAR R., LEHOCZKY J. P., "Priority inheritance protocols: an approach to real-time synchronization", *IEEE Transactions on Computers*, vol. 39, no. 9, pp. 1175–1185, 1990.

[ZAL 08] ZALILA B., Configuration et déploiement d'applications temps-réel réparties embarquées à l'aide d'un langage de description d'architecture, PhD Thesis, Ecole Nationale Supérieure des Télécommunications, November 2008.

List of Authors

Ludovic APVRILLE
Telecom ParisTech
France

Frédéric BONIOL
ONERA CdT
France

Etienne BORDE
Telecom ParisTech
France

Jean-Michel BRUEL
IRIT
France

Agusti CANALS
CS-SI
France

Joel CHAMPEAU
ENSTA-Bretagne
France

Jérôme DELATOUR
ESEO
France

Philippe DHAUSSY
ENSTA-Bretagne
France

Alain DOHET
DGA
France

Loïc FEJOZ
RTaW
France

Sébastien GÉRARD
CEA
France

Jérôme HUGUES
ISAE
France

Fabrice KORDON
UPMC
Paris
France

Brian LARSON
Research Associate
FDA Scholar in Residence
Kansas State University
USA

Philippe LEBLANC
IBM
France

Luka LE ROUX
ENSTA Bretagne
France

Chokri MRADHAI
CEA
France

Laurent PAUTET
Telecom ParisTech
France

Dominique POTIER
R&T Director
Pole System@tic
Paris
France

Ansgar RADERMACHER
CEA
France

Xavier RENAULT
Radio Protocol Embedded Digital
Systems Sector
Thales Communications &
Security
Paris
France

Thomas ROBERT
Telecom ParisTech
France

Jean-Charles ROGER
ENSTA-Bretagne
France

Pascal ROQUES
Independant
France

Pierre de SAQUI-SANNES
ISAE
France

François TERRIER
CEA
France

Béchir ZALILA
University of Sfax
Tunisia

Index